ANIMALS

PATRON

Dato' Seri Dr Mahathir Mohamad

SPONSORS

The Encyclopedia of Malaysia was made possible thanks to the generous and enlightened support of the following organizations:

DRB-HICOM GROUP

GEC-MARCONI PROJECTS
(MALAYSIA) SDN BHD

MALAYAN UNITED INDUSTRIES BERHAD

MALAYSIA NATIONAL
INSURANCE BERHAD

MINISTRY OF EDUCATION MALAYSIA

PERNAS INTERNATIONAL
HOLDINGS BERHAD

PETRONAS BERHAD

RENONG BERHAD

STAR PUBLICATIONS
(MALAYSIA) BERHAD

SUNGEIWAY GROUP

TENAGA NASIONAL BERHAD

UNITED OVERSEAS BANK GROUP

YAYASAN ALBUKHARY

YTL CORPORATION BERHAD

ACKNOWLEDGMENT

The Encyclopedia of Malaysia was first conceived by Editions Didier Millet and Datin Paduka Marina Mahathir. The Editorial Advisory Board, made up of distinguished figures drawn from academic and public life, was constituted in March 1994. The project was publicly announced in October that year, and eight months later the first sponsors were in place. By 1996, the structure of the content was agreed; later that year the appointment of Volume Editors and the commissioning of authors were substantially complete, and materials for the work were beginning to flow in. By late 1998, five volumes were completed for publication, and the remaining ten volumes fully commissioned and well under way.

The Publishers are grateful to the following for their contribution during the preparation of the first five volumes:
Dato' Seri Anwar Ibrahim,
who acted as Chairman of the Editorial Advisory Board;
and the following members of the Board:
Tan Sri Dato' Dr Ahmad Mustaffa Babjee
Prof. Dato' Dr Asmah Haji Omar
Puan Azah Aziz
Dr Peter M. Kedit
Dato' Dr T. Marimuthu
Tan Sri Dato' Dr Noordin Sopiee
Tan Sri Datuk Augustine S. H. Ong
Ms Patricia Regis
the late Tan Sri Zain Azraai
Datuk Datin Paduka Zakiah Hanum bt Abdul Hamid

SERIES EDITORIAL TEAM

PUBLISHER
Didier Millet

GENERAL MANAGER
Charles Orwin

PROJECT COORDINATOR
Marina Mahathir

EDITORIAL DIRECTOR
Timothy Auger

PROJECT MANAGER
Noor Azlina Yunus

EDITORIAL CONSULTANT
Peter Schoppert

EDITORS
Alice Chee
Chuah Guat Eng
Elaine Ee
Irene Khng
Jacinth Lee-Chan
Nolly Lim
Kay Lyons
Premilla Mohanlall
Wendy (Khadijah) Moore
Alysoun Owen
Amita Sarwal
Tan Hwee Koon
Philip Tatham
Sumitra Visvanathan

DESIGN DIRECTOR
Tan Seok Lui

DESIGNERS
Ahmad Puad bin Aziz
Lee Woon Hong
Theivanai A/P Nadaraju
Felicia Wong
Yong Yoke Lian

PRODUCTION MANAGER
Edmund Lam

VOLUME EDITORIAL TEAM

EDITOR
Kay Lyons

DESIGNER
Yong Yoke Lian

ILLUSTRATORS
Abdul Wahid bin Bulin
Anuar bin Abdul Rahim
Chai Kah Yune
Chu Min Foo
Ann Novello Hogarth
Peter Khang
Khor Choon Liang
Lee Sin Bee
Karen Phillipps
Sui Chen Choi
Tan Hong Yew
Wildlife Art Agency
Yeap Kok Chien

CONTRIBUTORS

Tan Sri Dato' Dr Ahmad Mustaffa Babjee
Consultant

Dr Stephen Ambu
Institute for Medical Research

Dr Edwin Bosi
Department of Wildlife, Sabah

Chan Chew Lun
Natural History Publications (Borneo) Sdn Bhd

Dr Chey Vun Khen
Forestry Department, Sabah

Arthur Y. C. Chung
Forestry Department, Sabah

Earl of Cranbrook
English Nature

Dr C. H. Diong
Nanyang Technological University, Singapore

Dr Durriyyah S. H. Adli
Universiti Malaya

Dr Ho Tze Ming
Institute for Medical Research

Jasmi bin Abdul
Department of Wildlife and National Parks

Dr Khoo Soo Ghee
Entomologist

Assoc. Prof. Dr Kiew Bong Heang
Universiti Malaya

Lee Han Lim
Institute for Medical Research

Dr Leh Moi Ung
Sarawak Museum

Dr Lim Boo Liat
Consultant

Mahedi Andau
Department of Wildlife, Sabah

Prof. Dr Mak Joon Wah
Universiti Putra Malaysia

Assoc. Prof. Mohamed Salleh Mohamedsaid
Universiti Kebangsaan Malaysia

Mohd Khan bin Momin Khan
Department of Wildlife and National Parks (retired)

Assoc. Prof. Dr Mohd Sofian Azirun
Universiti Malaya

Assoc. Prof. Dr Mohd Zakaria-Ismail
Universiti Malaya

Dr Peter K. L. Ng
National University of Singapore

Assoc. Prof. Dr Y. Norma-Rashid
Universiti Malaya

Pan Khang Aun
Department of Wildlife and National Parks

Siti Hawa bt Yatim
Department of Wildlife and National Parks

Sivananthan Elagupillay
Department of Wildlife and National Parks

Prof. Dr Yap Han Heng
Universiti Sains Malaysia

Prof. Dr Yong Hoi Sen
Universiti Malaya

THE ENCYCLOPEDIA OF
MALAYSIA

Volume 3

ANIMALS

Volume Editor
Prof. Dr Yong Hoi Sen

ARCHIPELAGO PRESS

Contents

Classification of Malaysian animals*

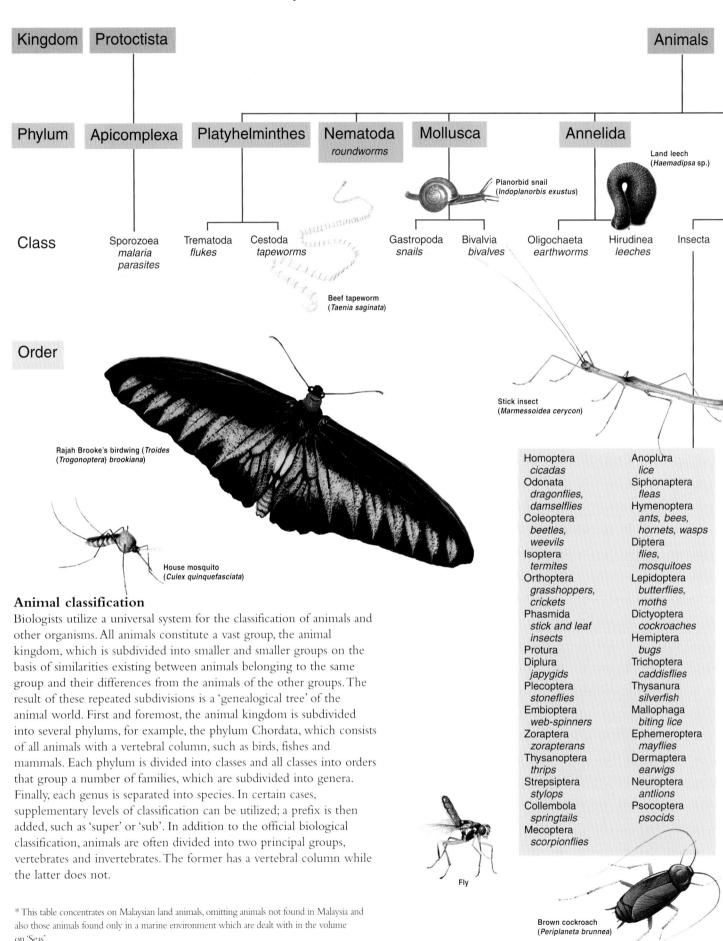

Kingdom | **Protoctista** | **Animals**

Phylum | **Apicomplexa** | **Platyhelminthes** | **Nematoda** *roundworms* | **Mollusca** | **Annelida**

Planorbid snail (*Indoplanorbis exustus*)

Land leech (*Haemadipsa* sp.)

Class | Sporozoea *malaria parasites* | Trematoda *flukes* | Cestoda *tapeworms* | Gastropoda *snails* | Bivalvia *bivalves* | Oligochaeta *earthworms* | Hirudinea *leeches* | Insecta

Beef tapeworm (*Taenia saginata*)

Order

Rajah Brooke's birdwing (*Troides* (*Trogonoptera*) *brookiana*)

Stick insect (*Marmessoidea cerycon*)

House mosquito (*Culex quinquefasciata*)

Homoptera *cicadas*	Anoplura *lice*
Odonata *dragonflies, damselflies*	Siphonaptera *fleas*
Coleoptera *beetles, weevils*	Hymenoptera *ants, bees, hornets, wasps*
Isoptera *termites*	Diptera *flies, mosquitoes*
Orthoptera *grasshoppers, crickets*	Lepidoptera *butterflies, moths*
Phasmida *stick and leaf insects*	Dictyoptera *cockroaches*
Protura	Hemiptera *bugs*
Diplura *japygids*	Trichoptera *caddisflies*
Plecoptera *stoneflies*	Thysanura *silverfish*
Embioptera *web-spinners*	Mallophaga *biting lice*
Zoraptera *zorapterans*	Ephemeroptera *mayflies*
Thysanoptera *thrips*	Dermaptera *earwigs*
Strepsiptera *stylops*	Neuroptera *antlions*
Collembola *springtails*	Psocoptera *psocids*
Mecoptera *scorpionflies*	

Animal classification

Biologists utilize a universal system for the classification of animals and other organisms. All animals constitute a vast group, the animal kingdom, which is subdivided into smaller and smaller groups on the basis of similarities existing between animals belonging to the same group and their differences from the animals of the other groups. The result of these repeated subdivisions is a 'genealogical tree' of the animal world. First and foremost, the animal kingdom is subdivided into several phylums, for example, the phylum Chordata, which consists of all animals with a vertebral column, such as birds, fishes and mammals. Each phylum is divided into classes and all classes into orders that group a number of families, which are subdivided into genera. Finally, each genus is separated into species. In certain cases, supplementary levels of classification can be utilized; a prefix is then added, such as 'super' or 'sub'. In addition to the official biological classification, animals are often divided into two principal groups, vertebrates and invertebrates. The former has a vertebral column while the latter does not.

Fly

Brown cockroach (*Periplaneta brunnea*)

* This table concentrates on Malaysian land animals, omitting animals not found in Malaysia and also those animals found only in a marine environment which are dealt with in the volume on 'Seas'.

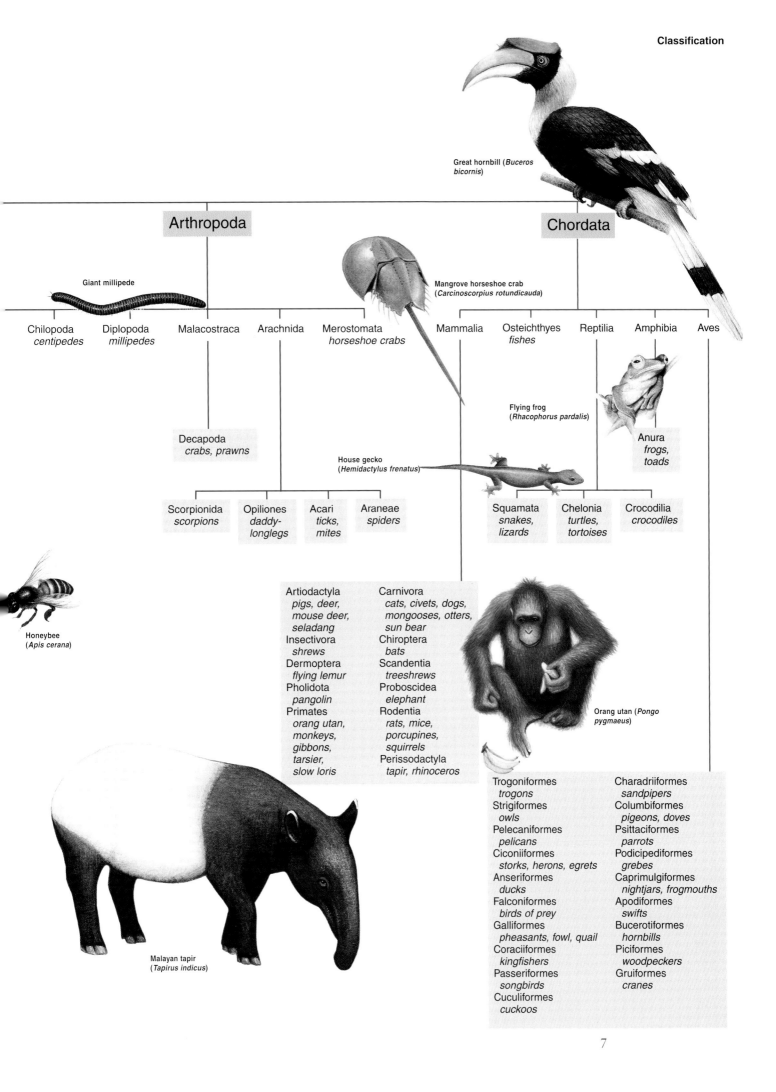

Great hornbill (*Buceros bicornis*)

Arthropoda

Giant millipede

Mangrove horseshoe crab
(*Carcinoscorpius rotundicauda*)

Chordata

| Chilopoda *centipedes* | Diplopoda *millipedes* | Malacostraca | Arachnida | Merostomata *horseshoe crabs* | Mammalia | Osteichthyes *fishes* | Reptilia | Amphibia | Aves |

Decapoda
crabs, prawns

House gecko
(*Hemidactylus frenatus*)

Flying frog
(*Rhacophorus pardalis*)

Anura
*frogs,
toads*

| Scorpionida *scorpions* | Opiliones *daddy-longlegs* | Acari *ticks, mites* | Araneae *spiders* |

| Squamata *snakes, lizards* | Chelonia *turtles, tortoises* | Crocodilia *crocodiles* |

Honeybee
(*Apis cerana*)

Artiodactyla
  *pigs, deer,
  mouse deer,
  seladang*
Insectivora
  shrews
Dermoptera
  flying lemur
Pholidota
  pangolin
Primates
  *orang utan,
  monkeys,
  gibbons,
  tarsier,
  slow loris*

Carnivora
  *cats, civets, dogs,
  mongooses, otters,
  sun bear*
Chiroptera
  bats
Scandentia
  treeshrews
Proboscidea
  elephant
Rodentia
  *rats, mice,
  porcupines,
  squirrels*
Perissodactyla
  tapir, rhinoceros

Orang utan (*Pongo pygmaeus*)

Malayan tapir
(*Tapirus indicus*)

Trogoniformes
  trogons
Strigiformes
  owls
Pelecaniformes
  pelicans
Ciconiiformes
  storks, herons, egrets
Anseriformes
  ducks
Falconiformes
  birds of prey
Galliformes
  pheasants, fowl, quail
Coraciiformes
  kingfishers
Passeriformes
  songbirds
Cuculiformes
  cuckoos

Charadriiformes
  sandpipers
Columbiformes
  pigeons, doves
Psittaciformes
  parrots
Podicipediformes
  grebes
Caprimulgiformes
  nightjars, frogmouths
Apodiformes
  swifts
Bucerotiformes
  hornbills
Piciformes
  woodpeckers
Gruiformes
  cranes

Introduction

The animal kingdom in the broadest sense covers 'all living organisms which cannot be classified as plants', though popularly the term 'animal' has acquired a more restricted meaning, namely mammals. In this volume, the term is used in its wider sense, as shown by the contents which represent all forms of animal life found in Malaysia.

A Malayan tiger (*Panthera tigris corbetti*), which belongs to the Indochinese subspecies, one of the five remaining subspecies; three others are already extinct.

The lime butterfly (*Papilio demoleus*) is common in the lowlands of Peninsular Malaysia, but is not found in Sabah or Sarawak.

Early Malaysia

Wild animals were plentiful in the little-disturbed lush green environment of early Malaysia. Thanks to the naturalists and archaeologists of both past and present, we are learning more and more each day of not only how the ancient inhabitants of the Malay Peninsula and Borneo (Sabah, Sarawak, Brunei and Kalimantan) lived, but also of the fauna and flora which existed then.

Most Malaysians would find it hard to imagine the abundance and variety of animals in our natural habitats a century ago. A chat with one's grandparents or great-grandparents, or a quick glance through the pages of the books written by late 19th-century travellers to Malaya such as Isabella Bird or William T. Hornaday, or an inspection of the early volumes of the *Malayan Nature Journal* would surprise residents of Kuala Lumpur. Herds of wild elephants and seladang roamed freely in the Batu Caves area and numerous crocodiles infested Sungei Buloh. Both these areas are only a few kilometres from the city centre.

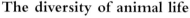

The yellow-breasted sunbird (*Nectarinia jugularis ornata*), one of Malaysia's tiny 'hummingbirds'.

The diversity of animal life

People living in rural areas of Malaysia may not understand the meaning of 'ecosystems' or 'biodiversity', but they are familiar with the great variety of animal species in the country. The diversity of animal life is spread out in adaptation to the varied environments, such as mangroves, mountainsides and rivers, as well as former mining land, primary and secondary forest and open fields. Some species are, however, versatile, adapting themselves to more than one environment, and are thus more commonly seen. This volume takes the reader to the different environments and consequently the diverse animal species of Malaysia. Although there is a growing number of city dwellers who are nature lovers, as development spreads across the country it is becoming more and more difficult for them to see the great variety of fauna without making some effort.

A pig-tailed macaque (*Macaca nemestrina*) and its baby among the green of a Malaysian forest. This is one of three macaque species in Malaysia.

Exploring nature

To see the diversity of animals present in just a small ecosystem, one needs only to visit a freshwater pond on the fringe of a city. There are animals at the bottom of the pond, as tiny as a full stop on this paper, and monitor lizards a metre long in the tall grasses around the water, in addition to a wide variety of fishes, shrimps, insects, tortoises, frogs, snakes, snails, spiders and wetland birds that can be seen if one knows how and where to look for them.

Malaysia's diverse animal life includes over 1,000 species of butterflies, more than 600 species of birds, 280 species of mammals, 140 species of snakes and 165 species of frogs and toads. Every year, new species of animals are still being identified or discovered.

Knowledge of animals' habitats and behaviour is essential in recognizing the great diversity of species, and why we must maintain

The long-tailed porcupine (*Trichys (Lipura) fasciculata*), the smallest of Malaysia's four porcupine species.

A stamp miniature featuring artwork of a number of Malaysian animal species in a rainforest setting. This stamp set was issued by Pos Malaysia in December 1996 with the purpose of making Malaysians more aware of their wildlife heritage.

their habitats in prime condition. All ecosystems are fragile. Even the smallest interference can destroy the diversity of species and result in single species domination. The relationships between various species of plants, animals, man and the physical and chemical elements must be understood. Disturbance in just one ecosystem can also affect all the other habitats. The diversity of Malaysian animals is a priceless gene bank that must be preserved by everyone; individuals, voluntary groups, corporate organizations and government agencies must all play their part.

Depletion of wildlife

Because of the richness of our fauna and flora, in the past trophy hunters and animal traders, as well as genuine scientific specimen collectors from the West and the East, journeyed to the Malay Peninsula and Borneo to trap, hunt or buy birds and insects, as well as mammals and reptiles, particularly tigers, elephants, seladang, orang utan, rhinoceroses and crocodiles. Fortunately, some of these specimens ended up in natural history museums or as specimens for scientific study.

A new era of hunting for fun, sport and exotic food began with the ownership of shotguns by planters. Wildlife depletion was further accelerated by illegal hunting and the popular habit of shooting animals for sport. The larger mammals, reptiles and birds suffered the greatest loss in numbers. Among such species were the barking deer and sambar deer, rhinoceros, tiger, mouse deer and crocodile as well as pheasants and pigeons. Poaching of wildlife is no longer the leading cause of depletion or extinction of species in Malaysia. The establishment of massive monocrop plantations, particularly of rubber and oil palm, and rapid industrialization have resulted in extensive loss of natural habitats for nearly all species. There have also been serious consequences from the widespread use of agricultural chemicals.

This volume

In the following chapters, a team of distinguished scientists and naturalists have assembled in one handy volume interesting and comprehensive information on the country's fauna. With growing numbers of Malaysians showing an interest in learning more about their natural heritage, this volume will more than serve their needs. It will also hopefully generate greater awareness for the need to protect our surviving fauna for the benefit and enjoyment of mankind and the well-being of our planet.

Juvenile estuarine crocodiles (*Crocodylus porosus*) in a crocodile farm, where they can be observed by visitors.

Game shooting was a popular pastime during the colonial era in Southeast Asia. Elephants, seladang and tigers were all eagerly hunted. Also, many tigers were shot because they attacked livestock as land was cleared for the establishment of plantations, which deprived them of their natural habitat.

Animals and people

The relationship between animals and the people of Malaysia is demonstrated in many different ways. The hunting of animals for food, the keeping of pets, the use of working animals in agriculture and transportation, the depiction of animals in currency and postage stamps, the use of animal products for decorative purposes—these are just a few of the examples of the importance which has long been placed on animals by the people of Malaysia.

The very tiny mouse deer (*kancil*) (*Tragulus javanicus*) is Sang Kancil, the hero of many Malay folk tales. Despite its small size, this wily animal is always the hero of such stories, often outwitting much larger forest animals, such as the elephant, tiger or crocodile.

Hematite cave drawings in the Tambun Caves near Ipoh, Perak, dating from the Neolithic period, which began about 4,500 years ago, include this drawing of a Malayan tapir with its very distinct black and white colouring. *INSET*: An adult Malayan tapir (*Tapirus indicus*).

Historical beginnings

Man's first relationship with animals was established out of necessity for food. The hunting instinct of the dog, for example, was used to help our own search for prey. In the earliest of times this worked both ways, but for most of the time the relationship has worked in the favour of people. A little later, when people began to recognize the intelligence and capability of animals, new relationships developed between them, but again they were to the benefit of people. They learned to domesticate other wild animals besides the dog and to use them for food, clothing, labour, transport and also for companionship.

Archaeological evidence

The evidence of these relationships is still being uncovered by archaeological excavations, and also from discoveries of man-made records from centuries before the art of writing was developed. Such evidence is seen in the many cave drawings which have been found in Peninsular Malaysia as well as in Sabah and Sarawak. Many wild animal species were an integral part of the survival of the early inhabitants of the island of Borneo, millions of years ago, and, later, of those who first inhabited the Malay Peninsula some 10,000 years ago. These people were brave, skilful trackers and gatherers, who could hunt down even the formidable seladang (gaur) (*Bos gaurus*) with their spears and stone adzes, not to claim the animals as trophies, but to provide a supply of food for their families. Even with the advent of modern weapons, these traditional hunting skills are still in use. Whether using modern or traditional weapons, some hunters still utilize dogs to assist in the chase.

In several places in the northern and central areas of Peninsular Malaysia, the floors of rock shelters and caves have yielded human remains, the tools they used, as well as remnants

A historical picture of a bird seller from whom town dwellers, unable to trap a forest bird themselves, could obtain a pet.

of the food, both plant and animal, that formed their diet. At Gua Cha and Gua Madu in Kelantan, for example, the remains of a wide variety of wild animals used for food have been identified, many dating back to the Mesolithic period. They included the shells and/or bones of river snails, turtles, rodents, monkeys, wild pigs, bears, barking and sambar deer, as well as the gigantic seladang.

Exploitation of animals

With the passage of time, people learned to further exploit animals for their benefit. Water buffaloes were used for the ploughing of paddy fields. Bullocks pulled carts carrying both goods and people. Elephants provided transport for the nobility, and brute strength for the movement of logs from forest to river. Macaques were trained to pluck coconuts. A variety of birds and mammals were kept as pets. In more recent times, dogs have been trained for police work, to assist in rescuing disaster victims as well as in the pursuit of criminals.

Love of animals

The affinity of Malaysians with animals is evident in *pantun* (Malay verse in rhyming couplets) and Malay proverbs, where the appropriate animals are used to convey subtle messages of love, disappointment or even revenge. The special status accorded to the tiny mouse deer (*kancil*) (*Tragulus javanicus*) as a cunning and intelligent creature is another clue to the regard Malaysians have for wildlife. Whilst honour and

Working animals

Animals have long been used in agricultural endeavours to ease the burden of work. Here are just a few of the useful roles for which people have trained animals.

1. A water buffalo pulls a cart laden with harvested oil palm bunches to be transported to a mill. Such carts are more suitable for the uneven, often muddy, terrain of an oil palm estate than motorized vehicles.

2. Bullock carts were used in the rural areas of Malaysia for many years before the widespread use of cars for the transportation of a wide range of goods and also passengers.

3. Now used mainly on ceremonial occasions, horses were traditionally used by the Bajau of Sabah for rounding up their livestock.

4. Pig-tailed macaques are trained to climb coconut palms and pick the fruit, following the instructions of their owner, who indicates whether unripe or ripe fruits are required.

respect are accorded to the mouse deer, traditionally there was great fear of some species, especially the crocodile and tiger.

The admiration and love of the Malays for some wildlife species were expressed by trapping them. In the kampongs, the animals most commonly kept were the peaceful dove (*merbuk*) (*Geopelia striata*) and the spotted-necked dove (*tekukur*) (*Streptopelia chinensis*). The keeping of these birds still continues, and competitions for the best songsters are held regularly in many places, especially in the northern states of Peninsular Malaysia. Mammals, too, were kept in captivity as objects of prestige, admiration or curiosity, but mainly by the sultans of the Malay states. From the records of 14th-century Chinese travellers, it is known that the Sultan of Kelantan was fond of barking deer (*kijang*) (*Muntiacus muntjak*) and had established a deer farm in the royal compound. Even the current coat of arms of the state of Kelantan is flanked by a pair of barking deer.

Animal products are often used in traditional ceremonial dress. Included in this costume are the feathers of hornbills, birds of great cultural significance to the people of Sarawak.

Brochures of two popular attractions in the Lake Gardens, the Kuala Lumpur Bird Park and the Kuala Lumpur Butterfly Park. These places not only provide an opportunity for visitors to see the many colourful birds and butterflies, but help to conserve these species by their breeding programmes.

Change in relationship

The relationship between animals and people will continue in the future, but the direction will change as people in Malaysia finally begin to realize that other living things have equal rights on this planet. People will serve their animal friends and repay the debts they owe the animals. This is already happening in the countries where poverty is almost unknown. Millions of dollars are being spent to save a fish, bird or mammal species and to preserve or restore the habitat.

Well-managed zoos and animal parks could be considered to be places where people serve animals, but in reality the relationship is mutual, as people derive great pleasure and satisfaction from seeing animals they do not have the chance to observe in their natural habitat. The role of zoos has evolved over the years from that of animal exhibition centres to become the scientific institutions for research, education, conservation and recreation that they are today. The attempts to breed the Sumatran rhinoceros (*badak sumbu*) (*Dicerorhinus sumatrensis*) at Melaka Zoo and the successful breeding of the milky stork (*burung upeh*) (*Mycteria cinerea*) at Zoo Negara are examples of research and conservation in action. It is envisaged that cageless and fenceless zoos will become more common as funds are made available for creating a more natural environment for animals.

These zoos will complement the protected areas which have been set aside to ensure animal survival in their natural habitat.

Medical research

Malaysian animals have been utilized in medical research programmes both at home and abroad.

In the past, many long-tailed macaques (*Macaca fascicularis*) were exported for use in experiments in overseas medical laboratories.

Within Malaysia, the mouse deer has been used in medical research at the Institute for Medical Research in Kuala Lumpur as well as at the universities.

Research is also being carried out on horseshoe crabs as their blue blood has been found to be a potent detector of gram-negative bacteria (the cause of most dangerous bacterial diseases in man). Test kits made from this blood can not only detect and analyse diseases but can also test if laboratory ware is free of bacterial contamination.

Animals and currency

Animals have long been depicted in the currency of Malaysia. In Sarawak and North Borneo (now Sabah), a miniature brass cannon with a buffalo head (1) was a very popular currency for barter trade. In Kedah, a figure of a metal fighting cockerel atop a series of rings (2) was minted during the reign of Sultan Mahmud Jiwa Zainal Abidin Muazzam Shah (1719–73). Tin ingots shaped to mark the first smelting of a mine took on animal shapes (3), including crocodiles, tortoises and insects, and became currency.

Status is still accorded to wildlife in Malaysia's paper currency. The logo of Bank Negara is the barking deer, and this animal appears in the left-hand corner of the reverse side of Malaysian bank notes. The value of Malaysian wildlife was further endorsed with its appearance on coins issued to commemorate wildlife conservation throughout the world. A 500-ringgit gold coin depicting the tapir (4), a 25-ringgit coin featuring the rhinoceros hornbill (5) and a 15-ringgit coin with the seladang were minted with the sponsorship of the World Wildlife Fund for Nature.

CANOPY

GROUND

MAMMALS

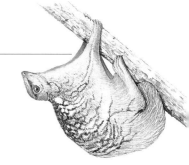

The flying lemur (*Cynocephalus variegatus*), which can be distinguished from flying squirrels by its lack of a tail outside the flying membrane.

Of the world's 4,600 species of living mammals (class Mammalia) (warm-blooded vertebrates which suckle their young), there are a total of 280 species in Malaysia, with 210 in Peninsular Malaysia and 220 in Sabah and Sarawak. The mammals of Malaysia range in size from the very large elephant to the tiniest of bats and mice. While it is the large mammals which tend to be the best known, the number of such species is small. It is the tiny animals which form the majority of the mammal species in Malaysia: bats comprise 40 per cent and rodents about another 30 per cent.

Mammals occupy virtually every kind of habitat available in Malaysia's rainforests—from bats, flying squirrels and the flying lemur in the tree tops to soil-burrowing mammals such as shrews, the moonrat and bamboo rats. At varied levels in the tree canopy are many more species, such as monkeys, cats and civets. Mammals living on the forest floor include the prey-eating tiger and plant-eating species such as the Sumatran rhinoceros and the Malayan tapir. Each species makes its home close to its source of food, which also influences the time at which each species is most active. For example, primates are mostly diurnal (move about during the day), while bats are mostly nocturnal (active at night). The tree squirrels are diurnal, but the flying squirrels are nocturnal.

While most Malaysian mammals are found in the lowlands, some are confined to high altitudes; others live at all elevations. There is also some division between the two regions of Malaysia. Many species occur in both regions, but some are found in only one. The tiger, for example, is found only in Peninsular Malaysia, while the orang utan, the western tarsier and the proboscis monkey are found only in Sabah and Sarawak.

Although related to other species in distant parts of the world, some Malaysian mammals are quite distinct from these relatives. The Malayan tapir is related to the tapir of South America, but has distinctive black and white coloration. The orang utan is related to the African apes, but is the only ape which builds nests, a practice also followed by the sun bear.

Some mammals are the heroes of legends; the mouse deer (Sang Kancil) is the most famous, with its uncanny ability to outwit even its largest opponent. Other mammals, such as wild pigs, provide food.

Mammals in the rainforest

A selection of Malaysia's mammals to illustrate their stratification in the rainforest and their division into diurnal and nocturnal animals.

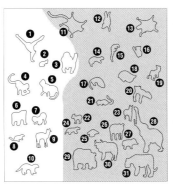

DAY

Canopy
1. White-handed gibbon (*Hylobates lar*)
2. Black giant squirrel (*Ratufa bicolor*)
3. Orang utan (*Pongo pygmaeus*)
4. Long-tailed macaque (*Macaca fascicularis*)
5. Silvered leaf monkey (*Presbytis cristata*)

Ground
6. Sun bear (*Helarctos malayanus*)
7. Pig-tailed macaque (*Macaca nemestrina*)
8. Three-striped ground squirrel (*Lariscus insgnis*)
9. Muntjak (*Muntiacus muntjak*)
10. Smooth otter (*Lutra perspicillata*)

NIGHT

Canopy
11. Red giant flying squirrel (*Petaurista petaurista*)
12. Flying fox (*Pteropus vampyrus*)
13. Flying lemur (*Cynocephalus variegatus*)
14. Common palm civet (*Paradoxurus hermaphroditus*)
15. Pangolin (*Manis javanica*)
16. Western tarsier (*Tarsius bancanus*)
17. Binturong (*Arctitis binturong*)
18. Slow loris (*Nycticebus coucang*)
19. Leopard cat (*Felis bengalensis*)
20. Leopard (*Panthera pardus*)
21. Pentail treeshrew (*Ptilocercus lowii*)

Ground
22. Mouse deer (*Tragulus javanicus*)
23. Sambar deer (*Cervus unicolor*)
24. Banded linsang (*Prionodon linsang*)
25. Malayan porcupine (*Hystrix brachyura*)
26. Malayan tapir (*Tapirus indicus*)
27. Wild pig (*Sus scrofa*)
28. Elephant (*Elephas maximus*)
29. Sumatran rhinoceros (*Dicerorhinus sumatrensis*)
30. Seladang (*Bos gaurus*)
31. Malayan tiger (*Panthera tigris corbetti*)

Evolution of the mammals of Sundaland

Mammals have a long history, but it is enough to go back only two million years, to the early Pleistocene era, to understand the origins and evolution of the present wild mammals of Sundaland, the region consisting of the Malay Peninsula, the large islands of Sumatra, Java and Borneo, as well as Bali and the other smaller islands of the area. Evidence is provided in fossils and through comparisons with living forms. It shows that successive invasions from continental Asia, followed by local evolution, have produced a distinctive regional land fauna.

Sabre-toothed cats, now extinct, existed from the time of the Oligocene epoch to the Pleistocene epoch. The name was taken from the long, blade-like canine teeth in the upper jaw. It is assumed that they killed by a slashing stab to the throat of their prey.

Mammal time line

25,000 years ago

Giant pangolin and its modern relative

All others are still extant today, but some of different size, and several over wider range.

800,000 years ago

Stegodonts
Southern mammoths
Hippopotamuses
Duboisia antelope
Early true elephant
(*Elephas hysudrindicus*)

Extant mammals no longer in region (e.g. hyena, hare)

Extant mammals still in region (e.g. orang utan, gibbons, pangolin, otters, wild dog, tiger, leopard, tapir, rhinoceroses, wild pigs, deer and cattle)

2 million years ago

Archidiskodont (southern mammoth) (*Elephas (Archidiskodon) planifrons*)
Sabre-toothed cat
Anthracothere (*Merycopotamus*)
Mastodon
Stegodont
Asian hippopotamus (*Hexaprotodon sivalensis*)
Duboisia antelope

Extant mammals still in region (e.g. monkeys, large cats, porcupines, pigs, wild cattle and deer)

Early mammals

With the cycle of polar cap freezing and melting, the sea levels in the area of the Sunda Shelf rose and fell. When the sea level rose, isolating land masses as islands, the animals on those islands were also isolated and, over time, evolved new characteristics and sometimes even new species. The selective effects of such geographical isolation operated even two million years ago.

Fossil mammal remains of about two million years ago found in Java represent animals which also occurred at that time in continental Asia. Most significant was the archidiskodont or southern mammoth (*Elephas (Archidiskodon) planifrons*). Now extinct, this primitive relative of the elephants ranged across Europe and Asia, and its fossil remains provide an invaluable date marker.

Other strange creatures that have since vanished include sabre-toothed cats, an anthracothere (*Merycopotamus*), a mastodon and a stegodont, two kinds of elephant-like mammals, a primitive Asian hippopotamus (*Hexaprotodon sivalensis*) and an endemic antelope (*Duboisia*).

Among the southern mammoths and the stegodont, dwarfed species evolved within Sundaland, two-thirds the size of their ancestors. Further east, the fossil remains of even smaller, half-sized stegodonts have been found on the Indonesian islands of Sulawesi, Flores and Timur.

In later fossil strata, about 800,000 years old, stegodonts, southern mammoths, hippopotamuses and *Duboisia* antelope occurred alongside an early true elephant (*Elephas hysudrindicus*), now also extinct. With these were extant mammals, some of which are still found in the region, for example, the orang utan, gibbons and the pangolin. Ground-dwelling, grazing and browsing mammals, as well as tree-dwellers and a wide variety of predatory carnivores, large and

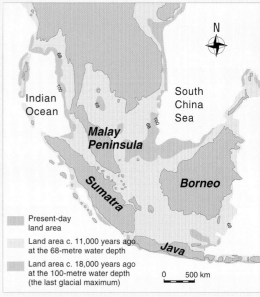

Indian Ocean
Malay Peninsula
South China Sea
Sumatra
Borneo
Java
N

Present-day land area

Land area c. 11,000 years ago at the 68-metre water depth

Land area c. 18,000 years ago at the 100-metre water depth (the last glacial maximum)

0   500 km

More than 300 million years ago, the earliest mammals evolved from reptile ancestors in a world very different from the present. About five million years ago, the processes of continental drift and land building finally formed the geographical region called 'Sundaland'.

For about 2.5 million years, the world's climate has oscillated between cold and warm periods in cycles of 40,000–100,000 years. During cold phases, global temperatures fell as low as 5 °C below present averages. The polar ice sheets expanded, locking up vast amounts of water. Sea levels fell throughout the world, exposing much of the Sunda Shelf as a large landmass connected to mainland Asia. It is likely that a continental type of climate prevailed across the wide central plain of Sundaland, with prolonged, alternating wet and dry seasons. Pollens in ancient peat deposits show associated changes in the vegetation, including the presence of pine trees at low elevations and increased areas of unforested land.

In the intervening warm phases, the seas rose and Sundaland was redivided into the Malay Peninsula and an archipelago of islands. A warm, equable and humid climate was restored, supporting rainforest vegetation in the lowlands.

small, suggest a varied habitat in which tracts of open, upland savannah alternated with extensive forests in the lowlands.

All the extant mammals of that period are (or were, in former times) also found on continental Asia, from where the original colonizers must have invaded Sundaland during one or more cold periods, when lowered sea levels meant animals could walk from continental Asia to the islands now surrounded by water. An early species of human, the apeman or pithecanthropus (*Homo erectus*) was among them, appearing in Java at about the same time as he did in southern China.

Not long before the last cool phase began about 25,000 years ago, one survivor from an earlier period was still present: a giant pangolin (*Manis palaeojavanica*) (which grew as long as 2.5 metres), existing alongside its modern relative, *Manis javanica*. All other mammals at that time were of kinds still extant today, although some differed in size from their modern descendants in Sundaland, and several occurred over wider ranges.

Local variants

Many of the scattered islets of the Sunda Shelf seas support populations of one or a few mammal species. The founding colonizers brought a limited gene pool, which was then subjected to intensive selective pressure in the constrained environment, resulting in local variants with minor physical differences.

Many small islands have been reached by only one of the two mouse deer found on the continental mainland and principal islands of Sundaland. Pulau Tioman, off the east coast of the Malay Peninsula, was apparently colonized by the greater mouse deer (*napuh*) (*Tragulus napu*), but the population has become dwarfed and is now hard to distinguish from the smaller lesser mouse deer (*kancil*) (*T. javanicus*).

For small mammals, the terrestrial environment can present an array of barriers to dispersal. Some which are now found only on mountains were more widespread during cool climatic phases. The ferret badger (*Melogale personata*) has differentiated into geographical subspecies on Mount Kinabalu (Sabah)

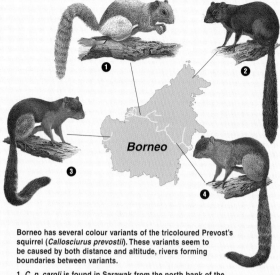

Borneo has several colour variants of the tricoloured Prevost's squirrel (*Callosciurus prevostii*). These variants seem to be caused by both distance and altitude, rivers forming boundaries between variants.

1. *C. p. caroli* is found in Sarawak from the north bank of the Rajang River to the Limbang and Lawas districts.

2. *C. p. pluto* is found in northern Borneo.

3. *C. p. borneoensis* is found from the Landak River, West Kalimantan, to the Saribas, Sarawak.

4. *C. p. atricapillus* is found across central Borneo from the Rajang River, Sarawak, in the north to the Mahakam River, East Kalimantan, in the south.

and in the Javan uplands of Indonesia.

In other cases, comparable selective pressures of high elevation habitats have led to the development of a superficial resemblance among unrelated species. Careful comparisons show that the three kinds of fluffy brown montane rats on the isolated high summits of Mount Leuser (Sumatra) (*Rattus hoogerwerfi*), Mount Kerinci (Sumatra) and Talakmau (*R. korinchi*) and Mount Kinabalu (Sabah) (*R. baluensis*) have evolved from separate ancestors.

Distance can be enough to perpetuate small inheritable differences. Genetic mixing among intermediate populations then creates a continuous, gradual variation in form, as among Bornean leaf monkeys of the grey group (*Presbytis hosei/everetti/frontata*) or the horse-tailed squirrel (*Sundasciurus hippurus*). Amongst the small lowland mammals, most variable is the Prevost's squirrel (*Callosciurus prevostii*). The distribution of distinctive colour variants of this species suggests that effective barriers include distance and altitude, as well as the lower stretches of rivers and estuarine creeks.

Changed forms

The rapid course of the final warming was probably marked by abrupt climate changes of decade or century duration. The highest sea level, around 5,000 years ago, was about 6 metres above the present level. Island areas enlarged and contracted with fluctuating sea levels. A warmer climate must have allowed the tree species of mixed dipterocarp rainforests to surge out of their lowland refuges, while vegetation types better adapted to colder, more seasonal conditions retreated to the montane zone or totally vanished.

Not all mammals were favoured by the return of closed high forest cover. Particularly disadvantaged were the hoofed animals which had grazed the savannah. Island populations were most vulnerable, once recolonization was prevented by the higher sea level. In Borneo, the larger of the Southeast Asian rhinoceroses, the Javan or one horned rhinoceros (*Rhinoceros sondaicus*), has become extinct within the last 12,000 years. The Malayan tapir (*Tapirus indicus*) survived there until at least 5,000 years ago but has since disappeared, as it has in Java. The wild dog (*Cuon alpinus*) also disappeared from Borneo.

The smaller, two-horned Sumatran rhinoceros (*Dicerorhinus sumatrensis*) survived, but underwent intriguing evolutionary change. Fossil remains from Sumatra and Borneo are slightly larger than modern specimens in the dimensions of their teeth, but more markedly so (in the range of 11–28 per cent) in the length of the bones of the limbs

and feet. Under selective pressures, a comparatively long-legged rhinoceros has evolved into a shorter, dumpier form. This trend, over a few thousand years, gives insight into the evolutionary processes that produced the dwarf southern mammoths and stegodonts of the remoter past.

Other cases of diminished size include wild cattle and pigs. Because teeth survive better than bony parts of the skeleton, in many cases it is the teeth that best demonstrate these changes. In orang utan, reduction has not been uniform, but relatively greater in the lateral incisors and rear molars. Among two species of forest rats, the third (last) molar has become relatively smaller with time. This trend parallels a widespread evolutionary tendency among Sundaland rodent species towards reduced or, in some, complete loss of the rear molars.

In their present distributions and taxonomic variations, the gibbons (*Hylobates* spp.) show how evolution has operated through the initial spread of an ancestral form, followed later by isolation, reinvasion and competitive exclusion among these exclusively arboreal mammals. The distribution of species of leaf monkeys (*Presbytis* spp.) has been interpreted similarly.

The archidiskodont was ancestral to all other forms of elephant that evolved in Sundaland. Compared with those of their descendants, their large cheek teeth were simpler and low-crowned. This tooth, seen in crown aspect (left) and side view (right), was found near Karang Jati, Java.

BELOW LEFT: The extinct giant pangolin of Sundaland (left) attained a total length of 2.5 metres, but otherwise resembled the modern species (right).

BELOW RIGHT: The tusks of the stegodont (left) were much longer than those of the modern elephant (right) and close together for much of their length, probably leaving no room between them for the trunk.

The orang utan

The orang utan (Pongo pygmaeus), the only great ape in Asia, is now found only in Borneo (including Sabah and Sarawak) and Sumatra, though fossils from Stone Age archaeological sites show that its original distribution stretched from China to Sulawesi. Even in Sabah and Sarawak, its present distribution is more restricted than in earlier times.

Distribution of orang utan in Malaysia

South China Sea

BRUNEI

Sabah

Sepilok Rehabilitation Centre

Semonggok Rehabilitation Centre

Sarawak

Kalimantan

N

□ Site of orang utan

The orang utan is often chosen as a mascot and included in logos. Here it is seen in the logo of the Sabah Tourism Promotion Corporation.

Names

Although the Malay name 'orang utan' ('forest man') has been universally adopted, this is not the name used in its home territory of Sabah and Sarawak. Names in Sabah include *kogiu*, *kahui* and *kisau*, while in Sarawak the orang utan is called *maias*.

Habits

A solitary arboreal primate, the orang utan mostly lives in rainforest below 1000 metres. The male lives an entirely solitary life, meeting with a female only for mating. The female travels with her baby until it is about six years old; by the time one offspring leaves her she probably has another baby to care for.

Only the young orang utan are at risk from predators, which are restricted to the clouded leopard and pythons. Thus, colonies are not needed for protection, nor for grooming—the orang utan groom themselves. The range of an adult orang utan is rarely more than 6 square kilometres; the daily range is less than 1 square kilometre.

Baby orang utan weigh about 1.5 kilograms, reaching 20–40 kilograms by 7–10 years. An adult female can weigh 35–50 kilograms, while a fully grown male is larger. Some are as heavy as 100 kilograms, with a height of 1.4 metres and a reach (both arms outstretched) of 2.4 metres. The life span of an orang utan is probably about 30 years. Their diet consists mostly (about 61 per cent) of fruits, but also includes leaves, bark, flowers and

Faces of two young orang utan (top), an adult female (with baby) (left) and an adult male (right). The black, square face of the adult male with its protruding cheek flanges of fibrous tissue is the obvious characteristic setting the Bornean subspecies apart from the Sumatran.

insects. One study showed that about 106 plant species are consumed. When selecting a forest site for rehabilitated animals, consideration must be given to how many of these favoured species are available in the area.

The orang utan is the only primate which makes a nest by breaking branches inward and adding smaller leafy branches for cushioning. It builds a new nest (sometimes several) every day. It is possible to estimate the orang utan population of an area from the number of nests sighted.

The orang utan moves slowly from tree to tree by oscillation. Upon grasping a branch of the next tree, it secures its hold with one leg before releasing the first branch.

Population

The present orang utan population in Sabah is estimated at 10,000–20,000, and in Sarawak fewer than 2,000. More than 1,400 orang utan have been removed from Borneo and Sumatra since the early 19th century for overseas zoos and research institutions. There are more than 1,000 in captivity around the world, and many are breeding successfully. The first captive births occurred in 1928—in the zoos of Berlin and Nuremburg in Germany and Philadelphia in USA—and more than 700 have since been born in captivity.

Although captive breeding is one of the options to save the orang utan from extinction, in the past there was a problem of interbreeding of Bornean (*Pongo pygmaeus pygmaeus*) and Sumatran (*Pongo pygmaeus abelii*) subspecies. This occurred because it is only in mature males that the difference between the two subspecies is readily discernible, as they have distinct facial features. However, chromosomal tests can now be used to identify the subspecies of an animal, and also hybrids, and so it is now possible to avoid further interbreeding.

Early rehabilitation

Surprisingly, it was not until the 1950s that orang utan were studied in their natural habitat. The method of the earlier naturalists, including Alfred Russel Wallace, was to kill them first! Thus they were able to study their physiology, but knew little about their life in the rainforest.

Barbara Harrisson with an orang utan in Bako National Park in 1964.

When the British administration in Sarawak began confiscating baby orang utan kept as pets, these animals became the responsibility of the curator of the Sarawak Museum, Tom Harrisson, who delegated the care of the orang utan to his wife, Barbara Harrisson. Receiving the first animal on Christmas Day 1956, Mrs Harrisson quickly developed a system for taking care of these animals in her own garden. Realizing that the ideal would be to prepare such animals for release into the forest, Mrs Harrisson made field trips into the rainforest to try to understand their way of life in the wild.

Rehabilitation was to be a long-term goal. At that time, there were no facilities for such a scheme. The animals 'adopted' by the Harrissons between 1956 and 1960 were considered too used to human company to be returned to the forest, and were sent to zoos, although only after careful vetting of the facilities and the treatment the animals would receive there. The first was sent to the USA; the others were sent to Europe.

However, from 1962 to 1965 Barbara Harrisson was able to try placing abandoned orang utan in the forest, at the Bako National Park. Unfortunately, for two reasons this plan did not succeed. The area was too small and too close to habitation, and there was no population of wild orang utan into which the released animals might integrate. The orang utan were later taken to the newly opened Sepilok Rehabilitation Centre.

Besides rehabilitation, a conservation measure in Sabah to save the orang utan from extinction is the translocation of animals restricted to pockets of forest as a result of deforestation. More than 200 animals have been moved to the wildlife forest reserve at Tabin, Lahad Datu, Sabah since 1993.

Rehabilitation centres

Orang utan displaced as a result of deforestation, babies of adult females shot by hunters, and those illegally kept as pets are all candidates for rehabilitation and return to their natural home, the rainforest.

The first rehabilitation centre, established at Sepilok on the east coast of Sabah, in 1964, has since become the largest such centre in the world. Another was later established at Semonggok in Sarawak. About 50 infant orang utan are received at the Sepilok centre each year. Some are surrendered voluntarily, while others are confiscated by an enforcement unit. More than 200 have been returned to the forest after rehabilitation.

After a 2-month quarantine period, the orang utan are gradually encouraged to learn the skills they would have learnt naturally if they were living in the forest. They learn to climb ropes to reach the feeding platform, where a seemingly monotonous diet of bananas and milk encourages them to look for other food (of a more traditional kind) in the trees. After some time, they build nests in the trees instead of sleeping in cages. Gradually they become less and less dependent on the centre.

Thousands of international travellers visit the centre at Sepilok each year to see the orang utan.

Monkeys and gibbons

All the three groups of Malaysian monkeys—the macaques, the leaf monkeys (langurs) as well as the very distinctive proboscis monkey of Sabah and Sarawak—belong to the family Cercopithecidae. There are also four small anthropoid apes, the gibbons, members of the family Hylobatidae. Two nocturnal arboreal primates are the slow loris and the only Malaysian forest 'gremlin', the western tarsier, which is found only in Sabah and Sarawak.

A trained pig-tailed macaque (*beruk*) (*Macaca nemestrina*) plucking coconuts. His owner stands at the foot of the palm giving instructions.

Silvered leaf monkeys (*Presbytis cristata*) at Bukit Melawati, Kuala Selangor.

Macaques

The long-tailed macaque (*kera*) (*Macaca fascicularis*), the commonest primate species found in Malaysia, is also found on offshore islands, such as Pulau Pinang, Pulau Langkawi and Pulau Tioman. As it can survive in patches of secondary vegetation near urban areas, it can be a nuisance. Its diet consists of leaves, shoots and small animals. A good swimmer, in coastal areas it hunts crabs, and so is sometimes known as the crab-eating macaque. It is a social animal, living in large groups. Although protected under the law, it can be kept as a pet if a licence is obtained. Prior to the 1980s, it was exported in large numbers for use in medical experiments in overseas laboratories.

Larger than the long-tailed macaque is the pig-tailed macaque (*beruk*) (*Macaca nemestrina*), a thickset animal which spends most of its time on the ground where it feeds on vegetation, fruits and small animals. Living alone or in small groups, it is best known as the monkey trained to pick coconuts.

Rarest is the stump-tailed macaque (*beruk kentoi*) (*Macaca speciosa*), which is only found in Perlis near the Malaysia–Thailand border. It is larger than the pig-tailed macaque, but also lives on the ground, only climbing trees when disturbed.

Nocturnal primates

The western tarsier (*kera hantu*) (*Tarsius bancanus*) (left) and the slow loris (*kongkang*) (*Nycticebus coucang*) (below) are both nocturnal animals. The tarsier, which lives in a small family group, jumps from tree to tree. Its bulging eyes allow it to spot its prey in the dark. Flattened discs on its fingers and toes help it cling to branches. The slow loris, on the other hand, gets a firm grip from the positioning of its thumb at right angles to the other digits. Much more slow-moving than the western tarsier, the slow loris allows its prey to approach before catching it with its forelimbs.

Leaf monkeys

The leaf monkeys (*lotong*) are distinguished from the macaques by their longer tail and longer hair and also their diet, which consists only of leaves. They are all gregarious, but vary in colouring and also habitat. The adult silvered leaf monkey (*Presbytis cristata*), which is dark metallic grey but has orange infants, is found along the west coast of Peninsular Malaysia, including Pulau Pinang, and also in coastal areas of Sabah and Sarawak.

The dusky leaf monkey (*Presbytis obscura*) is found in the lowland and hill forests of Peninsular Malaysia, including Pulau Pinang and Pulau Langkawi. Its body is greyish, with a distinct white circle around the eyes, and so it is also known as the spectacled leaf monkey.

Home for the banded leaf monkey (*Presbytis melalophos*) is the interior forests of Peninsular Malaysia. Adults have a greyish upper body, while the lower part is whitish. The young are pale grey with a dark cap and dorsal stripe.

Hose's langur or the grey leaf monkey (*Presbytis hosei*) and the red leaf monkey (*Presbytis rubicunda*) are only found in Sarawak and Sabah, while the white-fronted leaf monkey (*Presbytis frontata*) is restricted to central Sarawak.

Proboscis monkey

Found only on the island of Borneo, the proboscis monkey (*orang Belanda*) (*Nasalis larvatus*) lives at the river mouths of Sabah and Sarawak, and in peatswamp forests, feeding only on vegetation, including leaves, fruits and shoots. The male has a very prominent nose, in contrast to the female, but both sexes share the unusual grey, reddish brown and orange colouring. This monkey is a strong swimmer, often dropping off high branches into the water.

Gibbons

Gibbons (*ungka*) are characterized by swift movement through the forest canopy using only their very long arms to grasp branches—a process known as brachiation. They rarely descend to the forest floor, feeding in the canopy on leaves, fruits and insects. Gibbons live in small groups, with established territories.

The largest of the gibbons is the siamang (*Hylobates syndactylus*), an entirely black animal. It is confined to the interior forests of Perak, Kelantan, Pahang and Negeri Sembilan, usually living in small groups. An inflatable throat pouch enables it to make a distinctive call, louder than the other gibbons.

Peninsular Malaysia's lowland forests are home to the most abundant of the gibbons, the white-handed gibbon (*Hylobates lar*), whose hands and feet are of a much lighter shade than its brown body. Restricted to the north of Peninsular Malaysia is the agile gibbon (*Hylobates agilis*) while the Bornean gibbon (*Hylobates muelleri*) is found only in the lowland and hill forests of Sabah and Sarawak.

A selection of Malaysia's monkeys and gibbons in their typical habitat

1. White-handed gibbon
   (*Hylobates lar*)
2. Long-tailed macaque
   (*Macaca fascicularis*)
3. Siamang
   (*Hylobates syndactylus*)
4. Silvered leaf monkey
   (*Presbytis cristata*)
5. Proboscis monkey
   (*Nasalis larvatus*)
6. Pig-tailed macaque
   (*Macaca nemestrina*)

DISTRIBUTION OF MONKEYS AND GIBBONS

SPECIES	PENINSULAR MALAYSIA	SABAH AND SARAWAK
Macaques		
Long-tailed	✓	✓
Pig-tailed	✓	✓
Stump-tailed	✓	✗
Leaf monkeys		
Silvered	✓	✓
Dusky	✓	✗
Banded	✓	✓
Grey	✗	✓
Red	✗	✓
White-fronted	✗	✓
Proboscis monkey	✗	✓
Gibbons		
White-handed	✓	✗
Siamang	✓	✗
Agile	✓	✗
Bornean	✗	✓
Slow loris	✓	✓
Western tarsier	✗	✓

The elephant, the tapir and the rhinoceros

At first glance, the elephant, rhinoceros and tapir of the Malaysian rainforests seem to have little in common. However, a closer look reveals a number of similarities. All are ancient animals well camouflaged for their life in the rainforest, and all three animals are herbivores. The rhinoceros and tapir are both odd-toed ungulates belonging to the order Perissodactyla. The elephant and the tapir both have trunks, though the elephant's is much longer than that of the tapir.

In the 19th century, elephants were used for transport by the sultans and nobility of the northern states of the Malay Peninsula. In this photograph, probably taken about 1900, the female members of a royal family are posing on elephants. The man in the centre was possibly the elephant keeper.

The elephant

The elephant (*gajah*) species found in Malaysia is the Asian elephant (*Elephas maximus*); the other species is *Loxodonta africana*, the African elephant. In the 19th century, the elephant was numerous throughout Peninsular Malaysia. However, since the elephant requires a large home range and most of its habitat was cleared, it began to encroach into plantations, causing severe damage to crops. Planters saw the elephant as a threat to their crops, and turned to firearms and poisons to overcome this problem. As a result of inadequate legal protection, indiscriminate killing of elephants took place, and decades of such killing resulted in a serious population decline of the species. The status of the elephant became critical in the first quarter of the 20th century. In Peninsular Malaysia, wild elephants are now found only in small, scattered groups. In Sabah, wild elephants are found only in a very small area. This very limited distribution has led to speculation that they were introduced to Sabah, and that the existing wild elephants are descendants of imported captive elephants, perhaps tame animals given to the Sultan of Sulu by the British East India Company in 1750.

Peninsular Malaysia has an estimated population of about 1,000 elephants, mostly living in protected areas such as Taman Negara, the Krau Wildlife Reserve and the Endau-Rompin Park, as well as in areas of forest reserve and state land. Sabah also has an estimated elephant population of about 1,000.

Although the elephant is totally protected, the demand for ivory is a serious cause for concern. CITES (Convention on International Trade in Endangered Species) is being satisfactorily enforced in Malaysia, but smuggling is known to take place.

The tapir

Tapirs are now confined to Southeast Asia and South America, although fossil evidence shows they also once lived in Europe, continental Asia and North America. The Malayan tapir (*tenuk, cipan*) (*Tapirus indicus*) is the only species in Southeast Asia, and is the only tapir with the unusual black and white colouring, which provides excellent camouflage in the dappled forest environment. Now confined to Peninsular Malaysia, the tapir was found in Sarawak at least as recently as 8,000 years ago, as shown by fossils found in the Niah Caves.

While the tapir eats the young leaves and fruits of more than 115 plant species, 75 per cent of its diet comes from just 27 species. As a largely nocturnal animal, the tapir feeds mostly at night.

The tapir is a solitary animal. A male has a home range of about 12.75 square kilometres, but this overlaps with the home range of several other individuals. Resting areas are often found near rivers, but occasionally near the tops of ridges and spurs, away from water. Though the same bedding area may be used over a period of a week, different spots are selected. Dense cover is not required for resting—open places are often selected. The tapir does not wallow, but likes water, often entering rivers. A traditional Malay belief is that the tapir has the ability to walk along riverbeds, completely submerged, feeding on aquatic grasses.

Females generally produce one calf (occasionally two) every two years after a gestation period of about 401 days. Captive mothers have been observed to lick the newborn dry and eat the afterbirth. Captive born walk within one hour of birth.

A harmless animal, the tapir is naturally endowed

On an estate road in Sabah, a male elephant (*Elephas maximus*) protects his family from the danger of the intrepid photographer.

ESTIMATED POPULATION OF ELEPHANTS IN PENINSULAR MALAYSIA

STATE	NO. OF HERDS	NO. OF ANIMALS
Johor	17	138
Kedah	9	54
Kelantan	20	173
Negeri Sembilan	3	11
Pahang	41	205
Perak	10	130
Selangor	2	6
Terengganu	10	171
Taman Negara	7	120
Total	**119**	**1,008**

Relocation of elephants

The large-scale transformation of forests into oil palm and rubber plantations, as well as the spread of urban areas, has forced elephants into pockets of forest. To safeguard elephants, the Department of Wildlife and National Parks relocates such animals to safer areas, such as Taman Negara. This is done with the help of trained elephants. The animal is tranquillized for the journey, which is often a long one, sometimes

involving a raft ride across a lake. In order to keep track of the relocated elephants, the department has been joined by the Smithsonian Institution from the USA in a programme to attach transmitters to these elephants. The rough terrain and the heavy forest canopy make it difficult to monitor these transmitters from the ground or even from an aeroplane, and so satellite tracking has been used. Information about this programme can be found on the Internet where the Department of Wildlife and National Parks and the Smithsonian Institution both have homepages.

The adult Malayan tapir (*Tapirus indicus*) is dark brown or black with a white saddle from behind the front legs to the tail. A newly born tapir is dark brown with longitudinal white stripes and spots, which start to fade at the age of two months; adult coloration is attained by six months.

'How the tapir became black and white, and a vegetarian'

Once upon a time, says an Orang Asli tale, the tapir was a handsome fellow with a glossy black hide and a horn on the top of his head, and was as strong as a tiger. But he was vain and conceited, and refused to mix with the smaller, weaker animals. He demanded food and entertainment from all he visited. Finally, his tyrannical ways angered a poor group of rats and mice, who had barely enough food for themselves. Feeding such an enormous animal was impossible. So the chief, the bamboo rat, decided to teach the tapir a lesson. A potent drink was made from a special herb; this soon put the tapir to sleep. The bamboo rat and his friends quickly set to work. They filed down his teeth so he could no longer eat meat; they cut off his horn so he could not fight, and they painted his back with a permanent vegetable dye in order to brand him like a criminal. When the tapir awoke, he was astonished to see what had happened to him, and fled into the forest to hide his disgrace. Ever since, the tapir has been a shy, retiring vegetarian with a black and white body.

with protection from man. It is not hunted, probably because its meat is seldom consumed, but there have been cases of accidental shooting. The only known predator is the tiger, but attacks are rare.

Zoos worldwide are keen to exhibit the Malayan tapir. Already at least 116 are in zoos, and the growing demand can be detrimental to its survival. Though the tapir is totally protected, the ready market means wildlife traders are willing to pay good prices; this trade may be the main cause of the decline in tapir numbers.

The Sumatran rhinoceros (*Dicerorhinus sumatrensis*) wallows daily in mud to keep cool. The mud cover also protects the animal against flies. Wallows are indispensable to the rhino's good health.

The apparently smooth skin of the Sumatran rhinoceros (*Dicerorhinus sumatrensis*) becomes hairy once the animal is taken into captivity as it is no longer pushing its way through vegetation. Such an animal is referred to as a woolly rhino because its hair grows quite long.

The rhinoceros

The Sumatran rhinoceros (*badak sumbu*) (*Dicerorhinus sumatrensis*) is the smallest and most primitive of the five surviving rhinoceros species (two species in Africa, three in Asia). The very rare Javan rhinoceros (*Rhinoceros sondaicus*), which was previously also found in Malaysia, is now restricted to a small area on Java, Indonesia.

The larger of the two horns is on the tip of the muzzle; the second, above the eyes, is often no larger than a hump. The most striking feature is the two folds of skin encircling the trunk, just behind the forelegs and in front of the hind legs. The brownish grey skin appears smooth and granular, but close examination reveals small polygonal scales.

Driven back from the lowlands, the rhinoceros is now confined almost entirely to mountainous areas. It utilizes a wide area, often getting to the top of ridges as high as 1800 metres; tracks have been recorded as high as 3000 metres.

Many rhino trails appear like drains, 3–4 metres deep. Generations of rhinos have used these trails, and the animals seldom deviate from them. A rhino can travel up to 4.8 kilometres in a 24-hour period, and considerably more when disturbed.

Females bear one young. A newly born rhino weighs about 24 kilograms, quickly growing to an adult weight of 500–750 kilograms. Its height also increases rapidly—from 45 centimetres at birth to more than 100 centimetres at three years.

The Sumatran rhinoceros is the most endangered species in Malaysia, with only about 50 animals in Peninsular Malaysia, 20 in Sabah, and perhaps only one in Sarawak. Poaching is the main threat to its survival because of the high commercial value of rhino products. International trade in such products is driven by demand because of traditional beliefs that they have medicinal uses. Rhino horn is believed to reduce fever, rhino hide to cure skin diseases, and the stomach contents to relieve constipation. The monetary value of rhino products must be brought down to make poaching uneconomical, but the key to rhino conservation is for people to stop using rhino products.

GESTATION PERIOD (months)

Elephant	21–22
Tapir	13–14
Rhinoceros	16–17

What made them?

Many visitors to national parks are disappointed not to see any of the large forest animals. Proof of their presence, however, is provided by their tracks and piles of dung. Identification of the animals can be made from these clues.

Solutions: (1) Elephant (2) Tapir (3) Rhinoceros

The tiger and other cats

Malaysia has eight members of the Felidae family, all totally protected because of their rarity and vulnerability. The tiger, the panther and the golden cat are found only in Peninsular Malaysia, while the bay cat is found only in Sabah and Sarawak. Common to both areas are the clouded leopard, the marbled cat, the flat-headed cat and the leopard cat.

Postage stamps showing the clouded leopard (*Neofelis nebulosa*) issued by Pos Malaysia in 1995 to increase public awareness of this endangered animal.

The were-tiger

Fear of the tiger has given it an important place in superstitions of many Southeast Asian peoples. Traditionally, Malays have been reluctant to use the name of the tiger when talking about it. They prefer to use names such as 'Pak Belang' ('Mr Stripes'). Common to a number of societies is a belief in the were-tiger, a person who can change himself into a tiger. Such a belief has been used by leaders to instil fear into their followers, and also to collect contributions from those who do not wish to risk falling foul of the were-tiger. Some traditions, however, give the were-tiger a helpful personality. It is thought that a person who is a were-tiger can be identified by the absence of a cleft in his upper lip.

The tiger and panther

The Malaysian tiger (*harimau*) (*Panthera tigris corbetti*) belongs to the Indochinese subspecies, which ranges from Myanmar and Vietnam to Peninsular Malaysia. In 1992, the tiger population was estimated to be about 500. Most are in the major primary and secondary rainforest areas in the northeastern states of Kelantan, Terengganu and Pahang, including Taman Negara, which straddles these three states. There are also significant tiger populations in Perak and Johor. The panther (leopard) (*harimau kumbang/bintang*) (*Panthera pardus*) is thought to be rarer than the tiger.

Largest of the cats in Malaysia, an adult male tiger averages 2.62 metres in total body length (including tail) and a female, 2.38 metres. They can weigh up to 150 kilograms. The panther can reach 1.94 metres in length and weigh about 33 kilograms. It has a soft and dense coat, with a pattern of rosettes. Once considered to be separate species, the spotted panther and the black panther are now regarded as colour variations of the same species. The spotted panther is very rare; the last confirmed sighting was in Pahang in 1994.

Both the tiger and the panther are solitary species. An adult female tiger has a territorial range of at least 20 square kilometres, while a male's territory is larger, overlapping the territories of three or four females. At the age of two, a tiger cub establishes its own territory. The usual prey of both the tiger and the panther is wild pigs, although they do attack deer and other animals. Panther cubs start killing small prey by the time they are 6–7 months old. At the forest fringes, tigers and panthers prey upon livestock. In the past, this led to the shooting of rogue animals. However, since classification as totally protected species under the Protection of Wildlife Act, they are now driven back into the forest. In serious cases, the tiger or panther is caught and taken to the Melaka Zoo. Since 1984, more than 30 tiger cubs have been born in a captive breeding programme at the zoo.

The clouded leopard

Because of its arboreal and nocturnal nature, the clouded leopard (*harimau dahan*) (*Neofelis nebulosa*) is the least known of the big cats in Malaysia, but it is believed to live in most major forests. It can reach 1.65 metres in total body length and weigh up to 16 kilograms. Clouded leopards seem to live in pairs, hunting together for prey such as mouse deer, small wild pigs, monkeys, orang utan, porcupines,

rats and birds. Like panthers, they often spring on ground prey from overhanging branches. Cubs as young as three months old begin to kill their own prey.

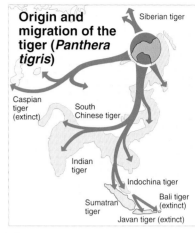

Origin and migration of the tiger (*Panthera tigris*)

Siberian tiger / Caspian tiger (extinct) / South Chinese tiger / Indian tiger / Indochina tiger / Sumatran tiger / Bali tiger (extinct) / Javan tiger (extinct)

The smaller cats

Very little is known of the distribution or ecology of the golden cat (*Captopuma* (*Felis*) *temmincki*), but it is believed to be distributed throughout Peninsular Malaysia. It can reach 86 centimetres in length, and weigh as much as 12 kilograms. Usually terrestrial, it can climb when necessary. Prey includes mouse deer, birds (especially pheasants), lizards and other small animals. The male is reported to play an active and tolerant role in rearing kittens, which weigh only about 250 grams at birth.

Probably the only wild cat still found in Kuala Lumpur is the leopard cat (*kucing batu*) (*Prionailurus* (*Felis*) *bengalensis*), the most widespread of the Malaysian cats. One entered a monitor lizard trap in the Lake Gardens in January 1995. In oil palm plantations, this cat helps to keep the rat population low. The leopard cat can reach 83 centimetres in total length, and weigh 4 kilograms.

The marbled cat (*Parado* (*Felis*) *marmorata*) is little known, probably due to its arboreal and nocturnal habits. It can reach 108 centimetres

ESTIMATED DISTRIBUTION OF TIGERS IN PENINSULAR MALAYSIA		
STATE	NO. OF LOCATIONS	NO. OF ANIMALS
Johor	26	38
Kedah	12	14
Kelantan	39	65
Negeri Sembilan	6	9
Pahang	48	84
Perak	51	81
Selangor	9	10
Terengganu	54	109
Taman Negara	27	81
Total	**272**	**491**

long, and weigh 2–3 kilograms. It preys on small animals such as mouse deer.

The flat-headed cat (*Captopuma* (*Felis*) *planiceps*) is widely distributed throughout Malaysia. In addition to the forested areas, it is known to be present in the Merbok mangrove forest in Kedah, the riverine habitat of Parit in Perak and in oil palm plantations. It is about the size of a domestic cat—about 66 centimetres long and 2.1 kilograms in weight. Captive animals seem fond of water, and like to eat fish, frogs, mice and meat.

Reported only from Sarawak and one unconfirmed sighting on Mount Kinabalu in Sabah, the bay cat (*Captopuma* (*Felis*) *badia*) is the least known of the Malaysian cats. It has a body length of about 1.12 metres.

Tiger cubs born in captivity in Malaysia, as part of the conservation programme to ensure the survival of the species.

Malaysia's cats in their rainforest home

1. Black panther (*Panthera pardus*)
2. Marbled cat (*Parado* (*Felis*) *marmorata*)
3. Clouded leopard (*Neofelis nebulosa*)
4. Spotted panther (*Panthera pardus*)
5. Leopard cat (*Prionailurus* (*Felis*) *bengalensis*)
6. Tiger (*Panthera tigris corbetti*)
7. Flat-headed cat (*Prionailurus* (*Felis*) *planiceps*)
8. Golden cat (*Captopuma* (*Felis*) *temmincki*)

23

The sun bear

The Malayan sun bear, or honey bear, the only bear in the Southeast Asian region, is the smallest of the world's seven bears. Its distribution ranges from northeast India, Thailand, Myanmar and southern China to Peninsular Malaysia, Sumatra and Borneo.

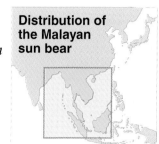

Distribution of the Malayan sun bear

1. The sun bear uses its long, sharp claws to tear open tree trunks to reach the contents of bees' nests.

2. A shattered tree trunk, the results of a sun bear's foraging.

3. The sun bear uses its long tongue to scoop up ants and termites from their nests.

Name

The Malayan sun bear (honey bear) (*beruang*) (*Helarctos malayanus*), which belongs to the family Ursidae, is a carnivore—a flesh-eating animal. The name 'honey bear' is related to its voracious appetite for honeycombs, while the name 'sun bear' is probably linked to the animal's fondness for basking in the sun on tree boles or on the ground, a practice usually found among cold-blooded animals.

Description

A powerful animal, when fully grown the sun bear weighs 27–65 kilograms and may stand up to 1.1–1.4 metres in height. Its short hair is uniformly black except for a white or pale buff V-shaped mark on the upper chest and a whitish muzzle, which is most conspicuous among cubs, becoming less pronounced as the bear matures. The head is large with small eyes and small, rounded ears. Its heavy body is carried on thickset limbs. Its tail is so short that it is scarcely seen. The paws are short and broad; each of the five toes has a long, curving and non-retractile claw.

In Malaysia, the species is widespread in forested areas at all altitudes but is nowhere abundant. The sun bear has poor hearing and sight; thus, it is solely dependent on smell to seek food. Even the facial whiskers, which are delicate organs for the sense of touch and are so well developed in many beasts of prey, are rudimentary and functionless in bears. Thus, dominated by their sense of smell, while dull in every other faculty, the sun bear lacks the alertness and decision of other animals whose lives are directed by a more balanced use of all the senses. The sun bear appears to be a primitive animal, yet in training it displays a high degree of intelligence, becoming a star performer in circuses and zoos.

Although most bears are nocturnal animals, the sun bear is more active during the day than at night. It is an excellent climber, using its powerful limbs, padded feet and powerful claws to go up large vertical boles. It can ascend the branchless trunk of a tall tree supported by its claws; thus, the presence of a bear can be detected by the characteristic gouged claw marks on the bark of trees. Sun bears sometimes travel about in pairs, but more often a solitary male or female with an occasional mother together with her young is encountered in the forest, according to the Orang Asli. Paired adults are usually encountered only during the mating season.

Nests

The sun bear usually sleeps in a tree nest. In 1940, A. H. Fetherstonhaugh described the nest-building process: *She would shin up a tree, climb out upon a limb until she reached a convenient fork where there were small leafy branches handy, and proceed to pull the twigs and leaves underneath her belly, her body along its length and all four legs hanging down. If overtaken by rain the procedure was the same and it was ludicrous to see her literally scuttle up the nearest tree and work against time to get a mat of leaves and twigs under her belly while leaving her back to the mercy of the elements; there she would stay with a look of patient misery on her face and not even hunger would get her down until the shower was over.*

Diet

Plants and insects, especially termites and earthworms, form the basis of the sun bear's diet. The bear uses its powerful claws to dig into termite nests, licking up the insects and eggs, while nests on the forest floor are opened by rooting with its nose and digging with its large paws. It also has a particular fondness for honey and bee larvae taken from hives high above the ground or hidden in tree hollows. The hives are knocked down to the ground and consumed there. During the process, the animal is often stung by adult bees, but as there are no reports of this animal succumbing to the poisonous stings of honeybees, it probably has a natural immunity to such poison. It is only when a bear is desperately hungry that it preys on vertebrate animals, such as ground birds, for example pheasants, and small mammals, such as civets, cats and rodents.

'How the sun bear lost its tail'

Once upon a time, a thin buffalo was accosted by a tiger, who threatened to eat him. However, the buffalo managed to get the tiger to agree to wait for seven days; by then, the buffalo promised, he would be fatter. On the seventh day, the buffalo met a crippled monkey who agreed to help him, and hopped on the buffalo's back. When the tiger appeared, the monkey munched on two eggplants, proclaiming loudly how good the tiger's head tasted. This frightened the tiger, who ran away and sought help from a bear. Both animals were afraid the other would run away, so they tied their tails together. But they were so frightened by the sight of the monkey still eating the 'tiger's head' that they both ran away, forgetting their tails were tied together. The bear's tail broke off, and never again did the sun bear have a tail.

THE MALAYAN SUN BEAR AND ITS COUSINS							
Common name	MALAYAN SUN BEAR	POLAR BEAR	BROWN BEAR	NORTH AMERICAN BLACK BEAR	ASIATIC BLACK BEAR	SLOTH BEAR	SPECTACLED BEAR
Scientific name	*Helarctos malayanus*	*Ursus maritimus*	*Ursus arctos*	*Ursus americanus*	*Selenarctos thibetanus*	*Melursus ursinus*	*Tremarctos ornatus*
Location	India to Indonesia	Arctic	USA, Canada, Europe	USA, Canada	Central Asia	India, Sri Lanka	Andes, Ecuador
Colour	black with pale V on chest and pale muzzle	white	brown	variable (white, brown, black)	black with white V on chest	black with white marking on chest	black with yellow marking on face
Length (adult male) (cm)	100–140	250–350	200	140–180	140–165	140–170	130–190
Weight (adult male) (kg)	27–65	500–600	150–375	57–272	90–115	127–145	80–125

Defences

Because the sun bear feeds on small creatures which burrow in the ground or hide under stones or the bark of trees, it does not pose serious competition to other carnivores, nor is it hunted by beasts of prey. The only likely enemies of the sun bear are a hungry tiger or panther.

The sun bear avoids man except when wounded or defending its young. Attacks on man are usually accidental. Being short-sighted and hard of hearing, a bear is likely to be surprised at close quarters, and if taken unawares will instinctively attack as a defence mechanism. In such circumstances, the sun bear is a really dangerous animal.

Nothing is known of the breeding pattern of the Malayan sun bear in nature, but in captive females usually one, and occasionally two, young are born at a time after a gestation period of about 95 days. The cubs are born blind and hairless, weighing about 300 grams. They stay with the mother for a year or more until nearly fully grown.

Conservation

Its medicinal and culinary value has historically made the sun bear a highly prized animal. The meat is regarded as a delicacy, and the gall bladder as having medicinal value in curing rheumatism and even as an aphrodisiac. However, the constant threat of man has decreased in Malaysia since the sun bear was declared a totally protected animal. Little is known of the status of the sun bear in Malaysia. However, because of habitat degradation due to development it is inevitable that numbers are declining.

Bears and dogs

From their appearance, it is difficult to believe that there is a relationship between dogs and bears. However, fossil remains indicate they are descendants of a common stock. The different modes of life adopted by the two animals led to the differences in structure now so apparent between dogs of the family Canidae and bears of the family Ursidae. The progenitors of the dogs became hunters and learned to capture prey by a swift and enduring chase. They developed slender, sinewy limbs and compact, short-clawed feet modified for swift movement on the ground. The progenitors of bears lived as bears live now, feeding on grasses, roots, herbs and insects, and occasionally meat. The search for such food does not require swift and agile movement, but rather legs built for climbing and digging, hence the bear's massive limbs which carry its heavy body up cliffs and trees, with in-turned paws to secure a better hold, and powerful claws to aid in climbing and digging.

The red dog, the sun bear's nearest relative

The distribution of the red dog (*serigala*) (*Cuon alpinus*) ranges from Central Asia, India and China, through to Java. In Peninsular Malaysia, it occurs in forested areas. It is distinguished from 'wild' domestic dogs by its colouring (rusty red with black tail), bushy tail, short, rounded ears, shorter legs and rather elongated feet with copious hair between the pads.

A comparison of the hind paw prints of a sun bear and a dog, its closest relative.

Canis familiaris

Helarctos malayanus

Otters, civets and mongooses

Malaysia has four otter species, twelve civet species and seven mongoose species—a larger number of mongooses than any other country. They are all medium-sized mammals of similar shape which belong to three separate families of the order Carnivora. The main distinguishing features between the three groups are their muzzles and feet, but there are also differences in habit—civets and mongooses are nocturnal, otters diurnal—as well as habitat.

The smooth otter (*Lutra perspicillata*) warming itself in the sun after a swim.

Otters

The otter's (*memerang*) close coat of waterproof fur, thick, muscular, slightly flattened tail and large, webbed feet are characteristics associated with aquatic behaviour.

The most common species, the smooth otter and the small-clawed otter, are distributed all over the country, mainly in the lowlands. All four species are protected by law, but little is known of the status of the other two species. The common otter, the largest of Malaysia's otters, has a rougher coat than the others, while the hairy otter is distinguished by a complete covering of short hairs on its nose.

All otters are highly territorial and are gregarious, moving in bands of 2–12 animals. Agricultural areas, especially paddy fields, as well as forests are home to otters.

Otters are powerful swimmers, but clumsy when moving about on land. Both males and females take part in digging the breeding den, which has an underwater entrance, carrying nesting materials into the den, as well as feeding the young.

Fish is the otter's main food, and its teeth are well adapted for dealing with this slippery prey. The presence of otters can be detected by piles of faeces (spraints), a mixture of crustaceans, molluscs and fish scales, deposited in open places near watercourses.

The Orang Asli believe that the meat of the otter, even if cooked well, has a strong odour and causes severe stomachache. Because of its stale, pungent, fishy odour, otter meat is generally also not favoured by other ethnic groups.

While otters are often viewed as pests because they invade aquaculture ponds, eating fish and also causing extensive damage, they also benefit the fisheries as they feed on frogs which eat fish eggs.

Civets

Civets (*musang*), nocturnal cat-like animals, are widespread throughout Malaysia, mostly restricted to forest habitats. Though they resemble the small forest cats in size, and some have patterns of spots or stripes, they are unrelated to cats. An obvious distinguishing feature is their pointed muzzle. The long black hair devoid of stripes or spots sets the binturong apart from the other species; also, its tail is

Distinguishing features of otters, civets and mongooses

Otters are distinguished by their broad muzzle, large, webbed feet and long toes; civets by their elongated head, pointed muzzle and short limbs; and mongooses by their sharp, pointed muzzle, tapering tail and slender legs.

At the left are shown the fore (left) and hind (right) footprints. Otters are distinguished by the web of skin between their toes, civets by the distinct toes arranged in a semicircle and mongooses by the small first digit.

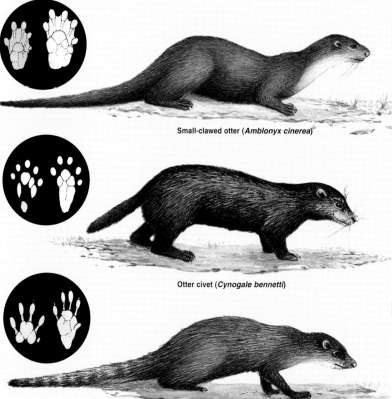

Small-clawed otter (*Amblonyx cinerea*)

Otter civet (*Cynogale bennetti*)

Javan mongoose (*Herpestes javanicus*)

COMMON NAME	SCIENTIFIC NAME	HABITAT
Otters (Family Mustelidae)		
Common otter	*Lutra lutra*	aquatic
Hairy otter	*Lutra sumatrana*	aquatic
Small-clawed otter	*Amblonyx cinerea*	aquatic
Smooth otter	*Lutra perspicillata*	aquatic
Civets (Family Viverridae)		
Subfamily Viverrenae		
Indian civet	*Viverra zibetha*	terrestrial
Large spotted civet	*Viverra megaspila*	terrestrial
Little civet	*Viverricula malaccensis*	terrestrial
Tenggalong	*Viverra tangalunga*	terrestrial
Subfamily Paradoxurinae		
Banded palm civet	*Hemigalus derbyanus*	terrestrial
Binturong (bear cat)	*Arctictis binturong*	arboreal
Common palm civet	*Paradoxurus hermaphroditus*	arboreal
Hose's civet	*Hemigalus hosei*	terrestrial
Linsang	*Prionodon linsang*	arboreal
Masked palm civet	*Paguma larvata*	arboreal
Otter civet	*Cynogale bennetti*	semi-aquatic
Three-striped palm civet	*Arctogalidia trivirgata*	arboreal
Mongooses (Family Herpestidae)		
Collared mongoose	*Herpestes semitorquatus*	terrestrial
Crab-eating mongoose	*Herpestes urva*	semi-aquatic
Hose's mongoose	*Herpestes hosei*	terrestrial
Indian grey mongoose	*Herpestes edwardsii*	terrestrial
Javan mongoose	*Herpestes javanicus*	terrestrial
Short-tailed mongoose	*Herpestes brachyurus*	terrestrial
Small Indian mongoose	*Herpestes auropunctatus*	terrestrial

prehensile, unlike the other civets. Only the linsang has retractable claws like a cat.

The otter civet, banded palm civet, linsang and binturong are totally protected; other species can be kept as pets (with a licence), but may not be killed for food.

All civet species except the linsang have scent glands. As only the binturong is not a silent animal, it is possible that these scent glands are used for communication. (The binturong, in contrast, howls loudly and utters low growls and hissing sounds.) Certainly the malodorous discharge, together with the striking colouring or marking of civets, provide an effective defence system.

Civet meat is believed to have aphrodisiac value, while the scent glands, mixed with herbal roots, have been used by some indigenous people as a love potion. Civets are of benefit to man as seed dispersers and controllers of plantation rats.

Mongooses

Mongooses (*cerpelai*) are small, slender-bodied mammals with head and body measurements of 25–60 centimetres. All seven Malaysian species are grey to reddish brown, but the crab-eating mongoose and collared mongoose are distinguished by their white and yellowish facial markings.

The most common is the short-tailed mongoose. While only one sighting of Hose's mongoose has ever been reported—in Sarawak in 1893—the crab-eating mongoose was first sighted in Malaysia in 1970. Snake charmers probably brought the Indian grey mongoose to Malaysia from India, where it is common. It is not found in Thailand, unlike the small Indian mongoose, which is common in the north of Peninsular Malaysia but rare elsewhere.

Despite being small, mongooses are rather bold and ferocious, often attacking and subduing animals much larger than themselves, such as the great argus pheasant. They also prey on other small vertebrates, such as rats and reptiles, while the crab-eating mongoose eats crabs, fish and other aquatic invertebrates. Though bold, mongooses are alert and wary, ready to scamper at the slightest alarm.

All species are restricted to forest areas except the short-tailed mongoose, which is also found in inhabited areas. They build nests in burrows under rocks or at the foot of trees.

Considered pests because of their attacks on poultry, mongooses are also beneficial because they control rats and mice in plantations.

DISTINGUISHING FEATURES OF MALAYSIAN CIVETS		
CIVET	**COLOUR**	**SIZE (cm)**
Tenggalong	all brownish to greyish	60–90
Indian civet	with black spots or stripes	60–90
Large spotted civet	on body and a ringed tail	60–90
Little civet	with white or black bands	50
Otter civet	uniformly brown	60–70
Hose's civet	uniformly brown	45–55
Banded palm civet	grey to reddish brown body with distinct dark bars across back and stripes on face; tail dark brown, banded at base	30–40
Binturong	long coarse hair, black mottled with brown	60–90
Common palm civet	grey with five distinct parallel dark stripes down back, spots on sides	40–60
Masked palm civet	body light to dark brown; white facial markings with black stripes and patches	40–60
Three-striped palm civet	light greyish brown to dark grey; three fine dark stripes from neck to base of tail	40–60
Linsang	pale yellow with black bars and spots; tail banded black and yellow	30–40

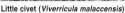
Little civet (*Viverricula malaccensis*)

Common palm civet (*Paradoxurus hermaphroditus*)

Masked palm civet (*Paguma larvata*)

Binturong (*Arctictis binturong*)

Three-striped palm civet (*Arctogalidia trivirgata*)

The mongoose and the cobra

Eye to eye and head to head
This shall end when one is dead
Turn for turn and twist for twist
Hah! The hooded Death has missed!

Rudyard Kipling's story of Rikki-tikki-tavi the mongoose and Nag the cobra is only one example of the traditional association of the mongoose and snakes, especially cobras. A factor in this association may be that the mongoose is known to be less sensitive to the venom of snakes than many other animals. However, it does not have total immunity, and a sufficiently large dose of venom will kill it.

Unlike most snakes, cobras do not simply slither away when threatened. The mongoose seems to appreciate the challenge of a cobra with forebody raised and hood expanded, ready to spray poison at any enemy which dares to come near. Confronted by a cobra, a mongoose may launch into a swift frontal attack (1), relying on speed to deliver a fatal blow. A more common strategy is a circumspect wait for an opportunity to attack; this comes when the cobra strikes and misses. At this point, the extended head is vulnerable to attack.

The mongoose often plays a deadly game by provoking an attack. It moves close to the cobra, pulling back just out of range at the critical moment (2). Its ability to raise its body hair, making the mongoose look twice as large and so closer to the snake causes the cobra to strike short (3). The mongoose takes advantage of the situation to make a final, fatal attack (4).

The seladang and wild pigs

While hunting of the seladang (gaur) (Bos gaurus), a very popular pastime among colonialists at the beginning of the 20th century, has long since ceased, Malaysia's two wild pigs, the common wild pig (Sus scrofa) and the bearded pig (Sus barbatus), are still popular targets of hunters. All three animals are even-toed ungulates belonging to the order Artiodactyla. The seladang is a member of the cattle family Bovidae; the pigs belong to the family Suidae.

A seladang (*Bos gaurus*), Malaysia's second largest mammal. *INSET*: A coin featuring a seladang issued by Bank Negara in conjunction with the World Wildlife Fund for Nature.

A new animal, the *selembu*, was born in 1984 when a wild seladang broke into a Veterinary Department cattle farm and mated with Friesian Sahiwal heifers.

Juvenile *banteng*

Banteng

The *banteng* (*Bos javanicus*), now found only in Sabah and Sarawak, is smaller than the seladang, and lacks the shoulder hump. Adult males are black, adult females and young animals (above) are brown. Both sexes have white 'stockings' on the lower legs and a distinctive white patch on the buttocks. The banteng is mainly nocturnal. Its diet consists mainly of grasses, and it visits natural mineral sources, including the sea. The banteng is mostly found in groups of about 8–10, with one mature male.

Although banteng are not domesticated in Malaysia, in Indonesia this has been done probably since prehistoric times; domesticated banteng are known as Bali cattle.

The seladang

The distribution of the species ranges from India and Myanmar to Thailand and Peninsular Malaysia, where it is found in the upper valleys of major rivers. It is not found in Sabah and Sarawak.

A massive black beast, the seladang has a shoulder height of 170–195 centimetres and weighs up to 900 kilograms. It has a high ridge on the back and white or yellowish white stockings on all four legs. The forehead is distinctly concave, with the frontal ridge thrown forward. The hair on the forehead and between the horns is pale grey to golden or white. Females and young animals are the same colour as old bulls, but newly born seladang are brown, not black, and can be differentiated from the young of common cattle by the dark stripe down the back. The structural feature of the horn and the colour pattern are the most striking characteristics.

The horns of the seladang grow outward and then curve inward, with the tip curved and a little forward. The tip of the horn is black, the middle third olive yellow, and the basal third dark olive brown, covered with thick corrugations. The extent of the corrugations and the sharpness of the tip of the horn depend on the age of the animal.

Essentially a lowland animal, the seladang's range in Peninsular Malaysia extends to the lower foothills of the main mountain ranges up to about 500 metres above sea level. Its habitat is the cool forest during the heat of the day and the open clearings and jungle edges during the night. It is a gregarious animal which stays in herds in specific areas, with herds sometimes overlapping in range.

A herd consists of adult cows, heifers and immature bulls, with a master bull as overlord. As young bulls mature, they leave the herd voluntarily or after conflict with the master, driven out to form their own herd. The domination of a herd is usually decided by combat, leading to terrific and bloody battles for supremacy. There are always solitary bulls who are anxious to acquire a herd for themselves.

As a grazer, the seladang is restricted to the vicinity of abandoned cultivations or natural clearings, such as the forest edge and open river banks. Unlike wild pigs, the seladang does not wallow. Grasses, such as paspalum and lalang, together with the foliage of a few forest trees, comprise its diet. The seladang has been known to inflict serious damage to plantation crops.

The gradual depopulation of this splendid beast is partly due to poachers, while the more significant factor is habitat destruction. Herds have shrunk from 30 or more 40 years ago to fewer than 10.

Tigers are the dominant predator, taking young animals in spite of the seladang's immediate and violent reaction to the presence of a tiger. An adult bull is a match for a tiger, as are cows in defence of their calves.

The Department of Wildlife and National Parks started a breeding programme 20 years ago, but this is a very slow process. What is more urgently needed is the total protection of the habitats inhabited by these beasts with very stringent enforcement. Unless this is done, the sustenance of the genetic pool of these beasts will be very bleak in the near future.

'Wit wins the day'

This story is taken from Walter Skeat's *Fables and Folktales from an Eastern Forest*, first published in 1901.

Pelanduk, the mouse deer, provoked two seladang (the wild bull of the clearing and the wild bull of the young bush) into a fight by telling each of them that the other had been making insulting remarks. With the mouse deer watching from a termite hill, the bull of the clearing killed the bull of the young bush. But the termites had eaten into Pelanduk's back, and so he could not move. He had to ask the bull of the clearing to knock down the termite mound, which the bull did before fleeing from the termites.

Pelanduk had begun to skin the seladang when Rimau, the tiger, appeared, wanting to share the meat. The mouse deer agreed, but sent Rimau off to fetch boughs to make a shelter as rain was falling. When the tiger returned, he found the mouse deer shivering. 'I am quivering with anticipation,' said Pelanduk when Rimau asked the reason for his shivering. Thinking that Pelanduk had designs on him, Rimau plunged into the river, leaving the meat to Pelanduk.

The wild pig

Malaysia's wild pig (*babi hutan*) is the common wild pig of Europe, northern Africa and Asia. It is a fairly large mammal with a shoulder height of 65–75 centimetres and weight of 75–200 kilograms. Females are considerably smaller than males. The young are born with a thick, distinctly patterned pelage (coat) but this coloration changes as they mature. In the adult, the body hair is black mixed with grey, white and sometimes brown. Whiskers are thick, turning grey with age. The ears are ovate and point backwards. Adult males have canines; both upper and lower sets curve outwards from the mouth.

The teeth of wild pigs are worn in the ears by this Sarawak tribesman.

The most propitious combination of habitats for the wild pig is apparently forest and cultivated fringes. Habitat selection is related to the availability of food, presence of water and thick vegetation cover. Tracks sighted in the forest are mostly near streams, rivers and swamps; they are also found in areas where trees are fruiting. During dry periods, wallows made by pigs are a common sight.

Wild pigs fight shoulder to shoulder, slashing each other with their tusks. Their shoulders have a thicker coat than their more vulnerable hindquarters.

Pigs of all age classes move in groups of varying sizes; 2–15 is common, but can be as many as 30. A family consists of the nursing cows and their litters. Male pigs leave the group when they become adult, returning only when the female is on heat. Adult males are solitary, and may fight for supremacy when they are attracted to a female on heat.

Both the wild pig and the bearded pig are omnivorous, eating various invertebrates, reptiles and small mammals, tubers, shoots, fungi, roots and fruits. They are pests of plantations, rooting up young seedlings of rubber trees and oil palms.

Wild pigs are a favourite prey of tigers and leopards, but man is their foremost predator. A large boar may hold its own against the big cats, but not man armed with powerful weapons. Deforestation is also leading to a decline in their numbers.

The bearded pig

The bearded pig (*babi jokut*), restricted to the Malay Peninsula, Borneo and Sumatra, is more abundant in Sabah and Sarawak than in Peninsular Malaysia, where it is confined to the southern regions.

Larger than the wild pig, with a shoulder height of about 90 centimetres and a weight of 57–120 kilograms, the bearded pig is distinguished by its pale colour, from reddish brown to buff white. However, the young are uniformly blackish. The most striking characteristic is the longer head with a 'beard' of bristles along the lower jaw and a fleshy protuberance with upward-pointing bristles above each side of the mouth.

Before giving birth, the bearded pig builds a massive nest (perhaps 2 metres wide and 1 metre high) where her litter will be well protected, in a dry area, often on top of a ridge. A depression is hollowed out, filled with dry earth and covered with a large pile of vegetation broken off nearby plants and placed with the broken ends pointing outwards. The nest may contain about 200 stems, some more than 3 metres long. The building of such a nest can thus have a devastating impact on the vegetation in the surrounding area.

The species is mostly active at night, but females with young are sometimes seen during the day. The bearded pig is suffering the same fate as the wild pig because of demand for its meat by certain ethnic groups. Unlike the wild pig, the meat of which is lean throughout the year, the meat of the bearded pig consists mostly of fat during the time of year when fruits are abundant in the forest. Although there is less information on the bearded pig, it should be considered more precarious in status than the wild pig as it is less abundant, and has a more restricted forest habitat. With deforestation and hunting of the species for food, more protective measures should be undertaken for the preservation of the bearded pig. It should not be placed in the same category of protected species as the wild pig, where hunting by licence is allowed all year round.

1. Bringing home a bearded pig from the hunt. *INSET*: Preparing a pig for cooking.
2. Punan Busang travelling by boat with their hunting dogs.
3. A Penan tribesman making darts for use with his blowpipe.

Bearded pig migrations

Periodically, large herds of several hundred individuals travel great distances in search of food, swimming across rivers, even to offshore islands, and climbing into mountain ranges. It is thought that such migrations may occur in those years when the *Shorea* trees have a heavy crop of *illipe* nuts, a favourite food of the bearded pig. In Peninsular Malaysia, such large herds have rarely been seen in recent years, indicating that the animal is fast depleting in numbers, due not only to habitat loss but also hunting pressures.

The wild pig (*Sus scrofa*) is a favourite target of hunters in Peninsular Malaysia.

Hunting

Though shotguns are now widely used for hunting, the traditional hunting weapons of the various indigenous peoples of Malaysia include blowpipes, bows and arrows, spears and traps. For hunting wild pigs, the Punan Busang of Sarawak use spears after their special hunting dogs have cornered an animal. The Iban set traps known as *peti*, using a forked sapling. One limb of the trap is embedded firmly in the ground; to the other is fastened a very sharp bamboo spear. When a pig enters the space between the limbs, a fibre stretched across the opening is displaced and the spear is released. As these traps are set near farms, it is necessary to warn people of their presence. This is done by placing beside the trap a *tuntun* (right), a carved figure that is also believed to attract animals to the trap.

Deer and mouse deer

Much more well known to many Malaysians than the three species of true deer are the greater and lesser mouse deer, the wily heroes of many folk tales. The Indian muntjak or barking deer has, in contrast, long been a favourite animal of Malay sultans and was in the past featured on gold coins. In more modern times, it is featured in the state crest of Kelantan as well as the logo of Bank Negara.

Kijang coins

Gold coins featuring a *kijang* (Indian muntjak) are attributed to a female ruler of Kelantan, Che Siti Wan Kembang. The date of her reign is uncertain—estimates range from the 14th to the 17th century. Legend says she was given a kijang by Arab traders. Kijang coins were in circulation in the northern Malay states and southern Thailand for centuries—possibly from 1400 until 1909, when Kelantan was placed under British administration.

Although these coins are called 'kijang coins', the animal featured on them, and on the Bank Negara logo adapted from them, has the head of a kijang but the feet, horns and tail of a bull, possibly taken from ancient Hindu coins circulated in the northern states of the Malay Peninsula.

'Sang Kancil and the crocodiles'

In Malay folklore, Sang Kancil, the tiny mouse deer, is a very cunning fellow who often outwits the bigger and more powerful forest animals. A typical story is that of Sang Kancil and the crocodiles.

Sang Kancil had spotted a rambutan tree loaded with ripe fruit on the other side of the river. So he told the crocodiles he had been asked by the king to count the number of crocodiles in the river. The crocodiles believed this story, and meekly lined up across the river. Counting as he hopped from one crocodile to another, Sang Kancil crossed the crocodile 'bridge' to feast on his favourite fruit.

Malaysian deer and mouse deer

Malaysia has three species of true deer belonging to the family Cervidae: the sambar deer (*rusa*) (*Cervus unicolor*), the Indian muntjak or barking deer (*kijang*) (*Muntiacus muntjak*) and the Bornean yellow muntjak (*M. atherodes*). There are also two mouse deer or chevrotains belonging to the family Tragulidae: the lesser mouse deer (*kancil*) (*Tragulus javanicus*) and the rarer greater mouse deer (*napuh*) (*T. napu*).

The main difference between the deer and the mouse deer is the stomach. While deer have the usual four-chambered stomach of a ruminant, mouse deer have a three-chambered stomach, lacking an omasum. Thus, mouse deer are considered the most primitive ruminants in the world. Mouse deer also lack the antlers which adorn the males of cervid deer. Instead of antlers, male mouse deer have extended canines which are capable of inflicting serious wounds. Similar extended canines are found in the male barking deer, which also has antlers.

Melaka state emblem

The state emblem of Melaka features the mouse deer to reflect the legend of the choice of the site for Melaka by the Sumatran-born prince, Parameswara, whose hunting dogs were so intimidated by a mouse deer that they fell into the river. A place where the weak could triumph over the strong was, Parameswara felt, an appropriate site for his new settlement.

Sambar deer

The distribution of the sambar deer is from India as far east as the Philippines. It is a fairly large, light brown to dark brown animal. Males have a short mane on the neck. There are two subspecies in

Deer horn carving

In Sarawak, high-quality hilts for Kayan and Kenyan *parang* (jungle knives) were traditionally made from deer antlers. The horn was worked with the small, all-purpose knife attached to the sheath of a parang. Horn carvers, highly respected craftsmen, made a sacrifice to the spirits and dedicated their work to a chosen supernatural helper. In addition to the fee for the carver, the buyer paid a separate fee for the guardian spirit. Deer horn was also used for hairpins (used by Kelabit men to keep their hair out of harm's way) and bottle stoppers for use in bottles of distilled rice spirit.

Malaysia, the Malayan sambar (*Cervus unicolor equinus* Cuvier), which is found in all states of Peninsular Malaysia except Pulau Pinang, and the Bornean sambar (*C. unicolor brookei* Hose), found only in Sabah and Sarawak. Unlike other deer, the sambar deer spends most of the day resting, and is active at night. While some are solitary animals, others live in small groups or large herds. Sambar deer live in the forest, but they are considered agricultural pests when they strip the bark off trees in plantations.

Muntjak

While the Indian muntjak is found in Sabah and Sarawak as well as in Peninsular Malaysia, the Bornean yellow muntjak is endemic to Borneo, and so in Malaysia is found only in Sabah and Sarawak. The name 'barking deer' is derived from the short, loud barking calls, which are used to scare away predators, and may also play a role in mating behaviour. Females bark and yap as often as males.

The Indian muntjak is reddish brown with darker feet and forehead, while the Bornean yellow muntjak is yellowish red. Newly born of both species have white spots, but lose these when they are about half adult size. The muntjak are much smaller than the sambar deer and are the only deer with frontal glands, a pair of slits on the face in line with the pedicel (base) of the antlers which are used for scent marking on the ground. Male Indian muntjak are adorned with short two-tined antlers on long, hair-covered pedicels; in Bornean muntjak, the antlers and pedicels are shorter. Their prominent tusk-like upper canines, which protrude outside the lower jaw, can move in their sockets and are often used in fights with other muntjak and for slashing predators.

The preferred diet of muntjak is fallen fruit, but they also eat herbs and young leaves. They live singly or in small groups.

Mouse deer

The greater and lesser mouse deer, tiny animals with very slender legs, are found from southern China to the Philippines. They are two of the world's four chevrotains, the smallest hoofed animals, and the only species in this region. The others are the African water chevrotain (*Hyemoschus aquaticus*) and the Indian chevrotain (*Moschiola meminna*). The lesser mouse deer is only about half the weight of the greater mouse deer, and the two species also differ in colouring. While the lesser mouse deer is reddish brown with distinct darkening of the nape and a single line of white from chin to chest, the greater mouse deer is greyish or orange buff without the distinctive darkening of the nape and with two bars of white. The fur of the greater mouse deer is not as soft as that of the lesser mouse deer, particularly around the shoulders, neck and nape.

Sebaceous glands just below the lower jaw are used for scent marking. Males have enlarged, tusk-like canines in the upper jaw, but do not have antlers. These teeth are often used for biting other mouse deer on the ears, neck and shoulders. Stamping, thumping or drumming on the ground with their hind feet, squeaking and fighting are all distinctive behaviour of the male mouse deer.

The two species of mouse deer share the same habitat, mostly upland forest, darting rapidly among the trees. They are nocturnal animals which live singly, in pairs or an adult with young, browsing on low bushes, grazing on ground vegetation and rooting in soil and surface litter; fallen fruits are also important in their diet.

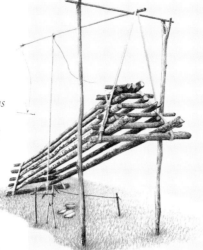

This type of trap, used by the Orang Asli to trap mouse deer, an important food item in their traditional diet, is operated by a trigger mechanism, enclosing the animal.

Conservation

The sambar deer has always been a popular choice of hunters because of its relatively large size and the quality of its meat, thus the need for conservation measures. The Department of Wildlife and National Parks has established three deer breeding programmes for sambar deer, at Sungkai Wildlife Reserve in Perak, at the Krau Wildlife Reserve in Pahang, and at Gua Musang in Kelantan. Some animals have already been released into the forests of Taman Negara.

The Institute for Medical Research, Kuala Lumpur, was the pioneer in the breeding of mouse deer. Universiti Putra Malaysia also has a breeding programme for mouse deer, with the hope of eventually returning animals to the forest.

Deer hunting

If you wish to wear bracelets and rings
Stretch out your two fore-feet

This is a deer hunter's charm quoted by Walter Skeat in his book *Malay Magic* (1900). He described the Malay method of deer hunting in great detail. After establishing the hiding place of the deer, elaborate preparations were made, beginning with a ceremony performed by the eldest of the hunters to induce the wood demons and earth demons not to interfere in the hunt. Without this ceremony, it was believed the expedition would not succeed. The equipment consisted of long coils of rattan rope, with nooses, also of rattan rope, attached at metre intervals. Near the deer's hiding place, a row of wooden stakes was driven into the ground. The rope was attached to the stakes, with the nooses hanging downwards. Two hunters, armed with knives, hid near the trap while the other hunters moved to the far side of the thicket. Moving towards the trap, shouting loudly, they startled the deer, which were quickly entangled in the nooses and killed by the hunters with knives.

The deer park in Kuala Lumpur's Lake Gardens displays Malaysian deer and mouse deer, as well as foreign species, allowing both city dwellers and foreign tourists a chance to view these shy forest animals.

A comparison of Malaysia's deer and mouse deer

There are vast differences in size between the smallest and largest of Malayia's deer and mouse deer species. Weights range from 2 kilograms of the tiny lesser mouse deer to 200 kilograms of the sambar deer. Shoulder heights also have a very wide range (20–120 centimetres).

The largest, the sambar deer, with a shoulder height of about 120 centimetres, weighs 120–200 kilograms. Much smaller is the muntjak (barking deer), with a shoulder height of 40–55 centimetres and weight of about 20 kilograms.

Only a fraction of the size of the true deer species are the two mouse deer. The lesser mouse deer is very tiny, with a shoulder height of 20 centimetres and weight of 2.0–2.5 kilograms. Twice as heavy, with a shoulder height of about 30 centimetres, is the greater mouse deer.

Sambar deer
(*Cervus unicolor*)

Muntjak
(*Muntiacus muntjak*)

Greater mouse deer
(*Tragulus napu*)

Lesser mouse deer
(*Tragulus javanicus*)

Shoulder height (cm)

120 —
100 —
80 —
60 —
40 —
20 —
0 —

The pangolin and porcupines

Malaysia has one species of pangolin and four porcupines. Although these nocturnal animals are totally unrelated, both have protective modified features as defensive weapons against predators. The pangolin belongs to the family Manidae of the order Pholidota and the porcupines to the family Hystricidae of the order Rodentia.

The Malayan pangolin (*Manis javanica*) in its native habitat

1. The pangolin uses its prehensile tail to keep a firm grip on a tree branch.
2. The baby pangolin suckles on its mother.
3. The female pangolin carries the young on its back, at the base of the tail.

The pangolin, when threatened, rolls itself up into a tight, impenetrable spiral to protect its non-scaly underside.

The pangolin

The Malayan pangolin (*tenggiling*) (*Manis javanica*) is a nocturnal animal found in plantations as well as primary and secondary forests, where it lives in a burrow. Its most distinctive characteristic is its armour of protective scales which covers its whole body with the exception of the underside, which is hairy. The pangolin looks more reptilian than mammalian, and its overlapping scales may be regarded as hairs or enormously enlarged spines, which are flattened with a few coarse, bristle-like hairs between them. It is light brown to olive brown with a head and body length of 35–50 centimetres, a tail of 20–38 centimetres, and a weight of 2–3 kilograms. Its feet are armed with large, powerful, curved and blunted claws for digging and tearing open the nests of ants and termites. The pangolin is toothless, but has a long, sticky tongue to lick up the ants or termites. Though terrestrial, it can also climb freely and does so in quest of tree ants. It climbs somewhat like a bear, gripping a bough tightly with the forelimbs and claws and, if necessary, with a curl of its tail, which is prehensile. Its diet consists primarily of termites and ants, as well as their eggs. It is particularly attracted by the large leaf nests of weaver ants (*Oecophylla smaragdina*) which hold swarms of adults, young and eggs.

Various legends and beliefs surround the pangolin. The most common story is of a pangolin playing 'possum'. According to this legend, a pangolin disturbs an ants' nest, after which it stretches itself flat on its belly with its scales fully erect, pretending to be dead. The ants, sensing a dead animal, swarm all over the body and get in between the scales, much to the delight of the pangolin. When it feels that the ants have got themselves thoroughly underneath the

This jacket belonging to an Orang Ulu shaman of Sarawak is covered with pangolin scales.

scales, it closes its scales, trapping the ants. The pangolin then trots to the nearest pool and submerges itself in the water. It raises its scales and laps up the ants as they float to the surface.

Pangolin scales are worn by some Orang Asli to drive away evil spirits. It is also believed that the meat and blood of the pangolin have both nutritional and healing properties. This belief endangers the species as people hunt the pangolin solely because of its purported medicinal values. The pangolin is beneficial to man as it helps in curtailing those ants and termites which are pests of both forest and plantation plants.

Bezoar stones

Malays have long believed that bezoar (*guliga*) stones (stones found in the bodies of animals) have magical properties. Although these stones can be found in a number of animals, it is thought that those from porcupines are the most powerful. Bezoar stones are worn as charms, or taken as medicine (scrapings, mixed with water). In the past, such stones were exported from Sarawak to India, where, in addition to use as medicine, they were used as an antidote to snake bite. Bezoar stones were sold by weight, using the gold scale; prices of stones varied according to quality and scarcity.

Porcupines

Porcupines (*landak*) belong to the rodent family Hystricidae. They are like gigantic rats, but are covered with spines (quills) which normally lie flat on the back, but can be raised so that they point out in all directions. The common porcupine (*Hystrix brachyura*) and long-tailed porcupine (*Trichys (Lipura) fasciculata*) are found in both Peninsular Malaysia and Sabah and Sarawak. The thick-spined porcupine (*Thecurus crassipinis*) is found only in Sabah and Sarawak, while the brush-tailed porcupine (*Atherurus macrourus*) is restricted to Peninsular Malaysia.

Porcupines are nocturnal animals and are terrestrial in habit. They are usually found in pairs, and live in crevices of rocks and in burrows. Porcupines feed on vegetable matter such as tubers, succulent stems and fruits, and are noted for taking objects back to their burrows. They are also known to gnaw at the bones of dead animals and even tusks of dead elephants, probably for the marrow of the bones and the pulp of the tusk.

A great deal of destruction is caused in gardens and plantations by the tunnelling of porcupines. In pineapple plantations, they eat the fruit and stamp on the plants. Similarly, they gnaw at the bark at the base of young rubber and forest trees, often destroying all the bark, resulting in the death of young trees. The animals are hunted as porcupine meat is considered a delicacy.

Malaysian porcupines

The brush-tailed and long-tailed porcupines are smaller species with a head and body length about half the size of the larger porcupines (25–40 centimetres) and a weight of 2–2.5 kilograms. The brush-tailed porcupine is slightly larger than the long-tailed porcupine. Although similar, they are distinguishable. The brush-tailed porcupine is brownish with flattened quills, interspersed with a few long, conspicuous quills. Its tail is about 20 centimetres long, ending in a tuft of curiously shaped hairs. Each hair looks rather like a string of flattened beads. The tuft, which is a dirty white colour, resembles a coarse Chinese paint brush; hence its name. In contrast, the long-tailed porcupine, the smallest of the four species, can easily be mistaken for a giant rat. The general colour is a rather muddy brown and it is distinguishable by its totally flattened quills without any interspersed long quills. Its tail is about 20 centimetres long and ends in a tuft of large, flattened hairs. This species, though once thought to be rare, is fairly common throughout Peninsular Malaysia.

Both the common and thick-spined porcupines are large species with a head and body length of 55–70 centimetres and a weight of 3–7 kilograms. They are not easily distinguishable from one another. Both have spines or quills 10–20 centimetres long, with specially modified short, hollow quills on their tail which rattle when shaken. The common porcupine is generally black with long, white quills with a black band towards the tip. In contrast, the thick-spined porcupine is generally dark brown with long, dark brown quills with a white tip and base.

Brush-tailed porcupine (*Atherurus macrourus*)

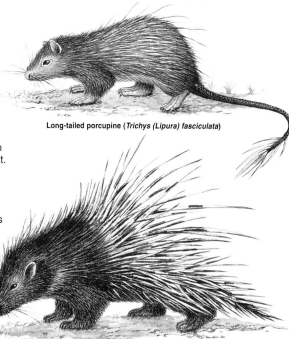

Long-tailed porcupine (*Trichys (Lipura) fasciculata*)

Common porcupine (*Hystrix brachyura*)

Thick-spined porcupine (*Thecurus crassipinis*)

Porcupine defences

When a porcupine is attacked, it (1) raises and rattles its quills, (2) charges backwards with incredible speed and (3) drives its erect quills into the limbs and body of its attacker. Faced with a bundle of sharp quills, many a foe is deterred. The belief that the quills can be discharged, like arrows, is not true, but the quills do come out easily.

❶

❷

❸

Bats

Bats (kelawar) (order Chiroptera) form the largest group of mammals in Malaysia, comprising 40 per cent of the 280 species of mammals. These include members of both suborders, Megachiroptera ('big bats'), the fruit bats, and Microchiroptera ('little bats'), the insectivorous bats.

When roosting, whether on a cave wall or in a tree, as here, bats hang upside down, using their feet to attach themselves firmly to their perch.

Fruit bats

Though belonging to the 'big' bat suborder, the 18 species of fruit bats recorded in Malaysia are of varied size. They range from the largest in the world, the giant fruit bat or flying fox (*keluang*) (*Pteropus vampyrus*) with a forearm length of over 18 centimetres and a wingspan of 1.5 metres to the smallest species with a forearm length of less than 4.5 centimetres. The most common of the fruit bats are the cave fruit bat (*Eonycteris spelaea*), commonly found in Batu Caves, Selangor, and the dog-faced fruit bats (*Cynopterus brachyotis* and *C. horsfieldi*) which roost in fruit trees and palms in residential gardens and occasionally in caves. Of the other fruit bats, which are primary forest inhabitants, the most common are the dusky fruit bat (*Penthetor lucasi*) and the spotted-wing fruit bat (*Balionycteris maculata*). Most of the fruit bats breed throughout the year, with peak periods coinciding with fruiting seasons.

Fruit bats (1) are distinguished from insectivorous bats (2) by their dog-like muzzle without noseleaves, large eyes, relatively small ears without a tragus and with a margin similar to that of man, a claw on the second digit, a very poorly developed interfemoral membrane and a very short tail or none at all.

Flight

Bats are the only mammals which can really fly. Their arms and hands are the framework of the wings, which are built on the usual pattern of the vertebrate forelimbs. The flexible movements of the skeletal structure adapt the wings to the twists and turns of flight, and adjust the surface to the changing current of air. Drawing the jointed finger bones reduces the wingspan, 'takes to sail' and instantly checks speed and momentum. With such flexibility at controlling its momentum, the wings of a bat are the most perfect flying organ created by nature.

Insectivorous bats

Most insectivorous bats have noseleaves, frills of skin around the nostrils lined with fine, sensitive hairs, which they use to detect danger. This gives them a distinct appearance from the fruit bats as do their large ears with margins which begin and end on the head, similar to a dog or cat. As befits their categorization as tiny bats, only a few of the 72 Malaysian species are large. Forearm lengths range from less than 3 centimetres to 9 centimetres. The species most commonly seen is the house bat (*Scotophilus kuhlii*), which roosts under the roofs of houses and in garden palms. Most species are forest dwellers; the commonest of these are the whiskered bat (*Myotis muricola*), greater and lesser flat-headed bats (*Tylonycteris robustula* and *T. pachypus*) and the trefoil horseshoe bat (*Rhinolophus trifoliatus*). The whiskered bat roosts in groups of 2–5 in tightly rolled central leaves of banana plants, while flat-headed bats roost inside the internodal sections

of bamboo, to which they gain access via vertical slits in the stem wall. Many species roost in limestone caves; the commonest of these is the diadem roundleaf bat (*Hipposideros diadema*).

Habits

All bat species are gregarious, living together in colonies. The local movements of these bats, the changes in feeding grounds and the nature of their food are largely influenced by the seasonal flowering and fruiting of trees which influence their abundance in a given area at one season and absence in another. Although fruit bats have good eyesight, they rely primarily on their keen sense of smell when in search of food.

A *Macroglossus* sp. fruit bat approaching a mangrove tree, *Sonneratia alba*. This bat is a forest-dwelling species which feeds on nectar and pollen.

The relationship of insectivorous bats with their environment is largely influenced by the creatures upon which they prey. High-flying bats, such as the lesser sheath-tailed bat (*Emballonura monticola*) and the black-bearded tomb bat (*Taphozous melanopogon*), come into contact with high-flying insects, while the smaller, less powerful Rhinolophid bats, such as the trefoil horseshoe bat (*Rhinolophus trifoliatus*) and the intermediate horseshoe bat (*R. affinis*), prey largely on moths, flies and other insects which keep to the lower level. Horseshoe bats hunt primarily in the forest, while some other species keep to the open country, to cultivated areas and around human dwellings; others hunt habitually over water. With their voracity and the myriad insects needed to sustain them, they are perhaps nature's most important check on nocturnal and crepuscular (those active at twilight) insect life.

Legends and beliefs

Bats seem frightening and mysterious, and no ghost story is complete without them. The belief that bats come out at night to suck human blood is derived from vampire folklore. The vampire bats of South America do suck the blood of animals, but Malaysia has only two false vampire bats, the Malayan false vampire bat (*Megaderma spasma*) and the Indian false vampire bat (*M. lyra*), which are not bloodsuckers.

It is believed that the meat of bats has medicinal value for people afflicted with asthma and rheumatism. In some Southeast Asian countries, bats are popularly hunted and sold as a delicacy.

Economic and medical importance

As habitual raiders of fruit plantations, fruit bats are considered economic pests. However, some nectar-feeding species are primary pollinators of mangrove forests and also aid in cross-pollination of fruit trees such as durian. Pollination of durian flowers is, however, mainly through wind currents and insects.

Pollinating species include the cave fruit bat (*Eonycteris spelaea*) and the common and hill long-tongued bats (*Macroglossus minimus* and *M. sobrinus*). As they reach into flowers with their long tongue, pollen collects on their head and shoulders; this pollen is later brushed off on other flowers, pollinating some. Any remaining pollen is later consumed by the bat as it grooms.

Fruit bats are also important seed dispersers. As they fly long distances from roosting sites to feeding grounds, ingested seeds are spread over a wide area. It has been found that the germination rate of some fig seeds increases after passing through the gut of a bat.

Insect-eating bats play an important role in pest control as they consume large quantities of insects carrying diseases and those which are crop pests.

Guano, the accumulation of dung and insect remains in caves by both groups of bats, offers sustenance and shelter to many insects and small creatures. It is also of importance to man as it is collected commercially as a highly prized fertilizer.

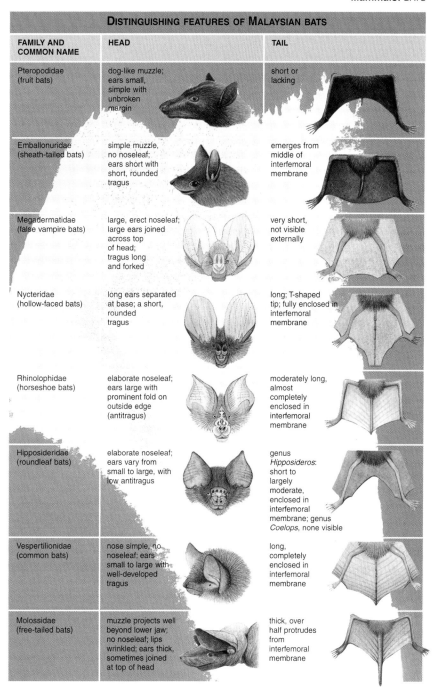

DISTINGUISHING FEATURES OF MALAYSIAN BATS

FAMILY AND COMMON NAME	HEAD	TAIL
Pteropodidae (fruit bats)	dog-like muzzle; ears small, simple with unbroken margin	short or lacking
Emballonuridae (sheath-tailed bats)	simple muzzle, no noseleaf; ears short with short, rounded tragus	emerges from middle of interfemoral membrane
Megadermatidae (false vampire bats)	large, erect noseleaf; large ears joined across top of head; tragus long and forked	very short, not visible externally
Nycteridae (hollow-faced bats)	long ears separated at base; a short, rounded tragus	long; T-shaped tip; fully enclosed in interfemoral membrane
Rhinolophidae (horseshoe bats)	elaborate noseleaf; ears large with prominent fold on outside edge (antitragus)	moderately long, almost completely enclosed in interfemoral membrane
Hipposideridae (roundleaf bats)	elaborate noseleaf; ears vary from small to large, with low antitragus	genus *Hipposideros*: short to largely moderate, enclosed in interfemoral membrane; genus *Coelops*, none visible
Vespertilionidae (common bats)	nose simple, no noseleaf; ears small to large with well-developed tragus	long, completely enclosed in interfemoral membrane
Molossidae (free-tailed bats)	muzzle projects well beyond lower jaw; no noseleaf; lips wrinkled; ears thick, sometimes joined at top of head	thick, over half protrudes from interfemoral membrane

Echolocation

Insectivorous bats have minute eyes, so have very poor eyesight. In the dark, they find their way about by echolocation, using a highly developed echo-apparatus—a radar system of their own. They emit supersonic sounds such as a faint clicking, buzzing noise and a high-pitched squeak. These sounds vibrate through the air and upon striking an object in their path are deflected back and instantly picked up by the bats. These warning echoes enable bats to locate and evade obstacles in their course, and play an essential part in guiding them during their search for food.

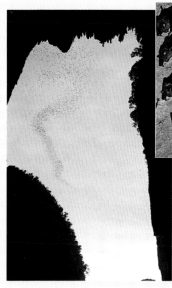

reflected sound waves
transmitted sound waves

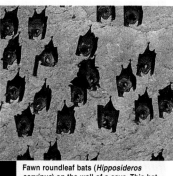

Fawn roundleaf bats (*Hipposideros cervinus*) on the wall of a cave. This bat species, which is found in caves in both Peninsular Malaysia and Sabah and Sarawak, roosts in very large colonies (up to 300,000). It feeds in the forest understorey.

LEFT: Wrinkled-lipped bats (*Tadarida plicata*) leaving their roost in the Deer Cave at the Mulu Caves, Sarawak. Large colonies of this species are also found in Sabah.

Squirrels and the flying lemur

As well as 26 species of diurnal tree and ground squirrels, Malaysia has 15 species of nocturnal flying squirrels, differentiated by their gliding membrane or patagium, which allows them to glide freely from tree to tree. Unrelated to the flying squirrels, the flying lemur (colugo) is easily confused with them because of its gliding ability.

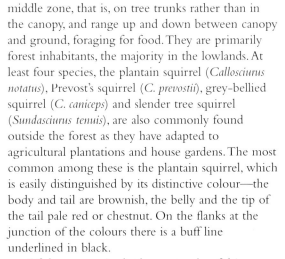

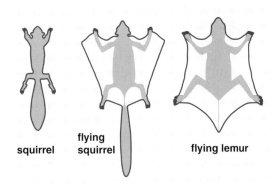

squirrel	flying squirrel	flying lemur

Tree and ground squirrels differ from flying squirrels in the lack of a gliding membrane which extends along each flank of the flying squirrel, joining the forelimbs to the hind limbs with the tail totally free. The gliding membrane of the flying lemur stretches from the angle of the jaw to the fingers and toes and to the very tip of the tail.

The very attractive Prevost's squirrel (*Callosciurus prevostii*), one of the medium squirrels which live mostly on tree trunks.

Squirrel nests
While flying squirrels roost in tree holes, some of the tree squirrels build globular nests of twigs.

Cross section of the nest

Malaysian squirrels
Squirrels (*tupai*) belong to the order Rodentia together with rats and mice, but are in a separate family, Sciuridae. The tree and ground squirrels belong to the subfamily Sciurinae, the flying squirrels to the subfamily Petauristinae. Of Malaysia's 26 species of diurnal squirrels, 20 are found in Sabah and Sarawak, and 14 in Peninsular Malaysia. They differ greatly in size: the pygmy squirrels are less than 10 centimetres in total length, the medium squirrels 18–50 centimetres, and the giant squirrels about 65 centimetres.

Tree and ground squirrels
The three giant squirrels are very colourful. Two species, the black giant squirrel (*Ratufa bicolor*) and cream-coloured squirrel (*R. affinis*), live in the canopy about 30 metres above ground level, descending only to cross gaps in the forest canopy; both inhabit lowland and foothill primary forests. The black giant squirrel, found only in Peninsular Malaysia, is easily recognized by its black body, huge black tail and orange or buff belly. In contrast, the cream-coloured squirrel has a brownish body with paler underparts. High in the forest canopy it uses twigs to build a neat, globular nest, from which it emerges well after dawn and to which it retires before dusk.

Largest of the giant squirrels is the tufted ground squirrel (*Rheithrosciurus macrotis*), confined to Sabah and Sarawak, which is dark brown with a blackish and a pale stripe on each side of the body; its belly is pale greyish buff. Its large, bushy, grizzled tail is held high when the squirrel is active or alert. The most distinctive character of this squirrel is the tufts of hair on the tips of the ears. When glimpsed briefly in the forest, its spectacular size, coloration and ground habit give the observer the impression of a larger mammal.

All the medium-sized tree squirrels live in the middle zone, that is, on tree trunks rather than in the canopy, and range up and down between canopy and ground, foraging for food. They are primarily forest inhabitants, the majority in the lowlands. At least four species, the plantain squirrel (*Callosciurus notatus*), Prevost's squirrel (*C. prevostii*), grey-bellied squirrel (*C. caniceps*) and slender tree squirrel (*Sundasciurus tenuis*), are also commonly found outside the forest as they have adapted to agricultural plantations and house gardens. The most common among these is the plantain squirrel, which is easily distinguished by its distinctive colour—the body and tail are brownish, the belly and the tip of the tail pale red or chestnut. On the flanks at the junction of the colours there is a buff line underlined in black.

Of the tree squirrels, the most colourful is Prevost's squirrel—the top of its head and tail are black, the belly and legs deep chestnut. A broad

Squirrel stratification
The various types of squirrels and flying squirrels live at varying levels, from the forest floor to the top of the forest canopy.

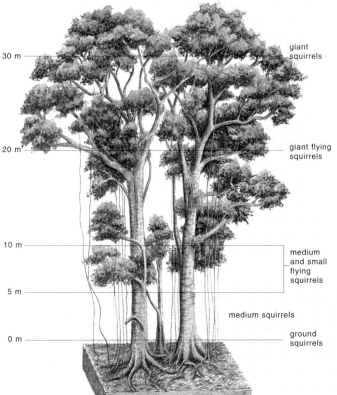

30 m — giant squirrels

20 m — giant flying squirrels

10 m

medium and small flying squirrels

5 m

medium squirrels

0 m — ground squirrels

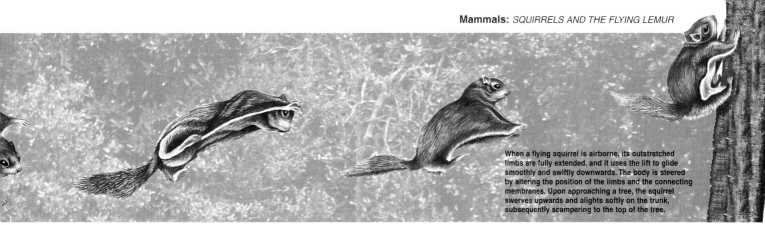

When a flying squirrel is airborne, its outstretched limbs are fully extended, and it uses the lift to glide smoothly and swiftly downwards. The body is steered by altering the position of the limbs and the connecting membranes. Upon approaching a tree, the squirrel swerves upwards and alights softly on the trunk, subsequently scampering to the top of the tree.

white stripe runs from the nose to the root of the tail on its underside. This very colourful squirrel is in great demand as a pet and as a zoo animal.

The ground species are easily distinguished by their short tail and elongated snout, and are not as colourful as the tree squirrels. Most striking of the ground squirrels is the shrew-faced ground squirrel (*Rhinosciurus laticaudatus*), which has largely adapted to feeding on insects. It has developed a long, sticky tongue, with a much longer snout to keep it in, and reduced incisor teeth which do not get in the way. The narrow, pointed head gives it the appearance of a tree shrew; hence its name. It has, however, found a use for its tail, which is carried fluffed out and at a high angle, like a giant catkin, presumably to distract attention from the body of the squirrel. An enemy taking a bite at such a morsel would get a mouthful of hair instead of squirrel.

Sabah and Sarawak have three pygmy squirrels which move between the trunk and canopy of forest trees. While Whitehead's pygmy squirrel (*Exilisciurus whiteheadi*) is known only from mountain forests, the black-eared pygmy squirrel (*Nannosciurus melanotis*) and the plain pygmy squirrel (*Exilisciurus exilis*) are found in both lowland and mountain forests. These two species can be distinguished by the pale stripe on the face of the black-eared pygmy squirrel. Their diet includes bark, lichens and small insects.

Flying squirrels

Flying squirrels are essentially nocturnal forest animals. Unlike tree squirrels, some of which make nests in the canopy, flying squirrels roost in tree holes which they have made themselves or which woodpeckers have abandoned. The circumference of the hole, its height and its size are associated with the different species that harbour in them. The giant flying squirrels usually roost in holes about 20 metres above ground in very large trees, while the medium and smaller sized squirrels' roosting holes are at heights of 5–10 metres in medium and smaller trees. Such stratification of roosts is associated with the different feeding habits of the different sized squirrels.

It is rare to be able to observe a flying squirrel when airborne. While sitting on a bough, the parachute is tucked close to the body and it resembles a giant tree squirrel (*Ratufa bicolor*).

The diet of the flying squirrel consists of fruits and nuts of various trees, while some species feed solely on green leaves and shoots. Some also eat bark and occasionally feed on insects. The larger species have also been known to feed on tree lizards and geckos.

Predators

The natural predators of squirrels are snakes, owls, eagles and small arboreal carnivores. But man is foremost of all. Squirrels are hunted for their meat, especially the larger species. Deforestation of their primary habitat has also contributed to population decimation of most of the forest squirrel species.

Feeding on a banana flower is the grey-bellied squirrel (*Callosciurus caniceps*), a medium squirrel which is also found outside forests, in plantations and gardens.

The flying lemur

The flying lemur or colugo (*kubung*) (*Cynocephalus variegatus*) is one of two members of the order Dermoptera. (The other member of the order is *C. volans* from the Philippines.) It is not a lemur (which is a primate), nor is it capable of true flight. Like the flying squirrel, it glides from tree to tree, but it is not only the flying membrane which distinguishes it from the flying squirrel. Its teeth are not of rodent form; the lower incisors are comb-shaped. The colour pattern varies greatly. Males are brighter and usually a shade of brown, whereas females are greyish. Both sexes have scattered white spots on the back. The underside of the gliding membrane is paler and is not spotted. The mottled coloration of this animal resembles the bark and trunk of trees, thus affording protection against predators.

An arboreal animal that is largely nocturnal, the flying lemur has been observed to be active in the early morning. It roosts in the holes of large trees about 15–18 metres above ground, and occasionally among palm fronds. When climbing or hanging upside down, its tail is folded under its abdomen and it rests spread-eagled against the trunk of trees or on the branches of coconut trees. Activity starts immediately after dusk.

The animal does not glide directly from the entrance of its hole, but crawls upwards for 1–2 metres before launching itself with outstretched limbs towards another tree. While in the air it floats like a kite and glides at approximately 60 degrees. Upon contact at its landing point, it raises its head slightly and with all four limbs still. It generally hangs upside down from smaller branches and twigs. In the forest it feeds primarily on young leaves and shoots, which it appears to pull down by means of its tongue; in coconut palms it feeds on flower buds.

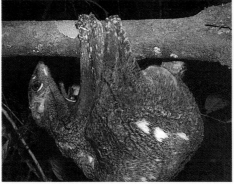
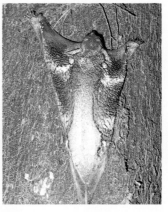

The flying lemur at rest. It hangs upside down from a branch (left) or lies spread-eagled against the tree trunk (right).

Shrews, rats and mice

Though it is only the few commensal species which are commonly seen, Malaysia's shrews, rats and mice make up about one-quarter of the total number of mammal species in the country; there are 9 shrew species, 10 treeshrew species, 27 rat species and 6 mouse species. Most are confined to the depths of the rainforest.

Shrews (right) differ from rats (centre) and mice (left) in having a long, pointed snout projecting considerably beyond the lower lip, a lack of incisor teeth and small, depressed ears. Mice are very similar in build and appearance to rats, but differ in size. The house mouse is a miniature replica of the house rat; the mouse has a head and body measurement of 5–8 centimetres in contrast to 15–20 centimetres for the rat.

Treeshrews (top), though very similar in appearance to squirrels (bottom), differ in having a long, pointed snout projecting far beyond the lower jaw.

Shrews

Malaysia's nine shrew species, belonging to the order Insectivora, are divided among three families: Erinaceidae with the two gymnure species, Talpidae with Malaysia's sole mole species, and Soricidae which contains six species of ground shrews. The most familiar of these is the largest, the house shrew (*cencurut rumah*) (*Suncus murinus*) (10–15 centimetres), as it is a commensal species (those dependent on man) often found in suburban gardens where its shrill, whistling chirp is heard at night. It is considered a nuisance as it enters houses, frequently defecating on the floor. The two other *Suncus* shrews are very much smaller. The pygmy shrew (*S. etruscus*), the smallest shrew in the world (3–5 centimetres), is found in Peninsular Malaysia as well as Sabah and Sarawak, but the black shrew (*S. ater*) is endemic to Sabah. The only known specimen was trapped on Mount Kinabalu.

The common forest shrew is the white-toothed shrew (*cencurut hutan*) (*Crocidura fuliginosa*), which lives on the forest floor, feeding entirely on insects. A similar species, though slightly smaller and lighter in colour, is the Sunda shrew (*Crocidura monticola*). Forest rivers are home to the Himalayan water shrew (*Chimarrogale himalayica*), distinguished by a fringe of short, stiff hairs on its feet.

Malaysia's only mole (*cencurut tanah*) (*Talpa micrura malayana*), found only in the Cameron Highlands, has a barrel-shaped body and very large claws on its front legs for digging speedily through soft earth to catch its prey (insects and earthworms) underground, where it lives in burrows.

Treeshrews

At various times classified as insectivores and primates, treeshrews are now considered a separate order, Scandentia, family Tupaiidae. Though similar in appearance to squirrels, treeshrews have a longer snout, pointed rather than chisel-shaped teeth and five clawed digits on each foot.

All are diurnal in habit except the pentail

The gymnures

The moonrat

The moonrat or gymnure (*tikus bulan*) (*Echinosorex gymnurus*), often given the role of king of the forest in Malaysian legends, is the largest of the Insectivora, the insect-eating mammals. It has a head and body length of about 35 centimetres. Its long snout somewhat resembles that of a pig.

When threatened, the moonrat makes a purring sound and emits a strong odour from its anal glands, perhaps the strongest of all the forest animals.

In Peninsular Malaysia, the moonrat has shaggy, black hair with white shoulders, naked tail and feet, and a black and white face. However, in Sabah and Sarawak it is almost entirely white, with a few sparse black hairs.

Strictly terrestrial, the moonrat ventures out of its forest home into nearby oil palm plantations. It also frequents forest streams in search of fish and aquatic vertebrates, which it eats in addition to insects.

The motif of the moonrat, which is popular in *pua kumbu* textiles of the Iban, is generally known as the white shrew motif.

Engraving of a pig-tailed shrew taken from John Whitehead's book, *Exploration of Mount Kina Balu*, published in 1893.

The pig-tailed shrew

The pig-tailed shrew or lesser gymnure (*tikus babi*) (*Hylomys suillus*) can be identified by its absurdly short, naked tail (only 1–2 centimetres long). Although related to the moonrat, it is very different in size and appearance. It is a brown animal about the size of a rat (with a head and body length of 12–15 centimetres) with a snout somewhat similar to a tapir's. Though it too emits a strong odour, it is not as malodorous as the moonrat. The pig-tailed shrew is found in forests at higher altitudes, sheltering in nests of leaves in hollows or under rocks, and feeding only on insects. When frightened, it freezes into a hunch and raises its long nose.

Watercolour of a bamboo rat with a clump of bamboo by a Chinese artist in the early period of British rule in the Malay Peninsula (about 1807).

treeshrew (*Ptilocercus lowii*), which is also distinguished by its naked tail with a quill-like end, and its entirely arboreal and strictly insectivorous habits. Despite their name, most of the treeshrews spend a great deal of time on the ground. All ten treeshrews are found in Sabah and Sarawak, but only three species in Peninsular Malaysia. Six species are endemic to Borneo.

The most abundant species is the common treeshrew (*tupai muncung besar*) (*Tupaia glis*), which is found at all elevations in all types of terrain from forests and plantations to house gardens. Though the common treeshrew spends much of its time on the ground, it is adept at climbing trees and sometimes jumps from one to another. It is a very vocal animal, a short tremulous whistling sound often heralding its presence. Alarm brings forth shrill, protracted cries as it dashes up a tree, followed by a series of little scolding grunts. At night it sleeps in a crudely constructed nest in the branch of a tree, a hollow log or the crevice of a boulder. This species can be an agricultural pest, particularly in cocoa plantations where it bites holes in the cocoa pods to reach the pith surrounding the seeds, and scatters the seeds all over the ground. It also damages developing rubber bud-shoots by cutting the succulent stem obliquely a few centimetres from the union.

Rats and mice

Both rats and mice belong to the order Rodentia, family Muridae. Most Malaysian species are forest dwellers; only a few depend on man for their survival.

Only two mouse species, the house mouse (*mencit rumah*) (*Mus musculus*) and the ricefield mouse (*mencit sawah*) (*Mus caroli*), are ground-dwelling species. The other species are all tiny, arboreal forest mice; three are endemic to Borneo. Only the common pencil-tailed tree mouse (*tikus buluh*) (*Chiropodomys gliroides*), which is found from India as far east as Bali, is a fairly common species. About 7–10 centimetres long,

it makes a nest of leaves inside bamboo stems, having first gnawed a hole in the bamboo, or in hollow tree branches.

The tree mice and tree rats differ from the ground-dwelling species in having, like monkeys, an opposable thumb with a flat nail instead of the usual sharp claw. Thus, they can secure a firmer grip.

Forest rats inhabit a wide range of ecological niches, from riparian and coastal swamp forests to montane forests. They are varied in size and in habit; some are strictly terrestrial, others partly arboreal, still others strictly arboreal. It is possible to divide the forest rats into two groups based on their fur type: ten soft-furred species and nine spiny-furred. The latter have short, coarse spines intermingled with the fur of the back. All the forest rats are omnivorous in habit, with a preference for fruits. None are serious agricultural pests, though some do damage crops in plantations close to forests.

Malaysia has seven species of commensal rats. Common urban species are the Norway rat (*Rattus norvegicus*) and the house rat (*R. rattus*); the Polynesian house rat (*R. exulans*) is found in both houses and fields. Field rats, such as the wood rat (*R. tiomanicus*), the ricefield rat (*R. argentiventer*) and the two bandicoots (*Bandicota indica* and *B. bengalensis*), damage food crops. However, it is not only economic losses from crop destruction and damage to building fixtures which must be considered. The health hazards from rodent-borne diseases are always cause for concern.

The bamboo rat

The large bamboo rat (*dekan*) (*Rhizomys sumatrensis*) and its sister species, the hoary bamboo rat (*dekan kelabu*) (*R. pruinosus*), are classified separately from the other rats, forming the family Rhizomyidae. Both animals are found from mainland Asia down to Peninsular Malaysia, but do not occur in Sabah or Sarawak. A thickset animal about 28–38 centimetres long, the large bamboo rat has short, coarse, brownish grey fur and a smooth, hairless tail. Its strong claws are used to dig burrows under the bamboo clumps which provide its food. When threatened, it displays its large, broad incisor teeth to deter enemies.

The lesser treeshrew (*Tupaia minor*), one of the three species found in Peninsular Malaysia.

The pencil-tailed tree mouse (*Chiropodomys gliroides*), one of the tiny arboreal forest mice.

The wood rat (*Rattus tiomanicus*), which causes damage to plantations and gardens.

The common treeshrew (*Tupaia glis*), the most abundant species, found in many habitats.

The long-tailed giant rat (*Leopoldamys sabanus*), a soft-furred forest rat.

The grey tree rat (*Lenothrix canus*) (with young), an uncommon soft-furred forest rat.

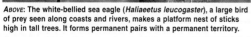

ABOVE: The white-bellied sea eagle (*Haliaeetus leucogaster*), a large bird of prey seen along coasts and rivers, makes a platform nest of sticks high in tall trees. It forms permanent pairs with a permanent territory.

RIGHT: One of Malaysia's numerous waterbirds, the purple heron (*Ardea purpurea*) is often seen standing motionless at the water's edge in swamps and mangroves, looking for prey.

FAR RIGHT: The white-breasted kingfisher (*Halcyon smyrnensis*) is the most frequently seen of the 15 kingfisher species found in Malaysia.

BIRDS

Worldwide, there are 8,600 species of birds (class Aves) divided into 27 orders comprising 155 families. Of this large number, Malaysia has 624 species of birds belonging to 78 families. Only 426 are resident species. The others are migrants or vagrants which use the East Asian Flyway to migrate from the northern hemisphere at the beginning of winter. Some winter in Malaysia, flying north again at the approach of the northern summer; others stop for only a short time before continuing their long journey.

The rhinoceros hornbill (*Buceros rhinoceros*), whose colourful, upturned casque makes it the most distinctive hornbill species.

In this book, only a small number of these numerous species can be included. One of the most significant orders of Malaysian birds is the hornbills, of which the country has 10 species. Although often associated with Sarawak because of the strong cultural significance to that state, Peninsular Malaysia has more species of these large forest birds. Also closely associated with Sabah and Sarawak are the economically important edible birds' nest swiftlets. These birds are also found in Peninsular Malaysia, but it is from Sabah and Sarawak that collection of these nests has become valuable business. Trade in birds' nests, a delicacy in Chinese cuisine, began several centuries ago as barter trade with China, but the processed nests are now exported to many parts of the world.

Other birds of cultural importance are some of the songbirds, by far the largest order of birds. Malays have traditionally kept two dove species, the zebra or peaceful dove (*merbuk*) (*Geopelia striata*) and the spotted-necked dove (*tekukur*) (*Streptopelia chinensis*), as cage birds because of their melodious voice. In recent years, songbird contests for these two species, as well as others, such as the white-rumped shama (*murai batu*) (*Copsychus malabaricus*) and the red-whiskered bulbul (*merbah telinga merah*) (*Pycnonotus jocosus*), have become very popular. Birds which constantly win prizes in such contests can become very valuable. Mynas are also well liked as cage birds because they are good mimickers, especially the hill myna (*tiong mas*) (*Gracula religiosa*), which can even mimic human speech, and the common myna (*tiong gembala kerbau*) (*Acridotheres tristis*).

There are two distinct types of birds of prey, the magnificent eagles and hawks of the order Falconiformes, which are active during the day, and the nocturnal birds—owls of the order Strigiformes and frogmouths and nightjars of the order Caprimulgiformes—which sleep by day and hunt at night.

Deep in the rainforest are found the birds with many 'eyes' on their wings, the pheasants. These are the only Malaysian birds which clear, and jealously guard, dancing grounds for performances to attract a female. One species, the Javanese green peafowl (*merak*) (*Pavo muticus*), has been hunted to extinction. However, there are plans to reintroduce it to its natural habitat, using captive birds.

Waterbirds are found not only along the coast, but also at inland bodies of water, both natural and man-made. This group includes a wide variety of species, both residents and migrants. Among these species are herons, storks, egrets and waterhens as well as ducks and geese. One very rare species, the milky stork (*Mycteria cinerea*) has been successfully bred in captivity at Zoo Negara.

y tiny songbirds with long, slender beaks obtaining nectar from deep inside flowers, birds hover in a manner somewhat similar ummingbirds (which are not found in this on). As with many birds, the male is much e colourful than the female. Here a ple-throated sunbird (*Nectarinia sperata*) eeding its young in a pear-shaped nest le of grass and leaves typical of the birds.

collared scops owl (*Otus lempiji*) is a very nocturnal bird of Malaysia, weighting 65 ns or less. It makes its nest in a hollow tree, re it lays 2–5 white, unmarked eggs.

Birds of prey

Malaysia has 41 species of diurnal birds of prey, also called raptors, belonging to the order Falconiformes, some predators and others scavengers. These birds of prey occupy all types of habitat, from the mud flats of the coastal areas to the mountains. Some species are residents, some are migrants, while a few are vagrants, stopping only to feed and rest before continuing their migratory journey.

Characteristics

Birds of prey are characterized by their hooked, down-curved bills and strong, clappered, clawed feet, which are used for clutching prey and for tearing flesh. In most species, the female is larger than the male. They occur singly, in pairs or in family groups of 3–4, but never in large groups. Occupying the highest position of the food chain in the animal kingdom, their prey ranges from the smallest bees to chickens. Most nest in trees, making their nests of dried branches and twigs arranged in a large pile. However, the black-thighed falconet (*lang rajawali*) (*Microhierax fringillarius*) nests in cavities of tall trees.

With the exception of the black-thighed falconet, which belongs to the family Falconidae, all the Malaysian birds of prey belong to the family Accipitridae, comprising several groups of birds.

Langkawi

Pulau Langkawi is, by legend, named after a brown eagle (*lang kawi*), commonly seen on the island and adopted as its symbol. A 12-metre-high sculpture of the bird has been erected at the entrance to the harbour where it is seen by tourists arriving by boat from the mainland, as well as those arriving by air from overseas.

Among the birds of prey seen in Malaysia are both resident and migrant species. In their natural habitat are a number of these birds.

1. Marsh harrier (*Circus aeruginosus*)
2. Crested honey buzzard (*Pernis ptilorhyncus*)
3. Changeable hawk-eagle (*Spizaetus cirrhatus*)
4. Black-thighed falconet (*Microhierax fringillarius*)
5. Brahminy kite (*Haliastur indus*)
6. Black kite (*Milvus migrans*)
7. Black-shouldered kite (*Elanus caeruleus*)
8. White-bellied sea-eagle (*Haliaeetus leucogaster*)
9. Osprey (*Pandion haliaetus*)
10. Black baza (*Aviceda leuphotes*)
11. Crested serpent-eagle (*Spilornis cheela*)
12. Blyth's hawk-eagle (*Spizaetus alboniger*)

Resident birds of prey

The contrast of white head, neck and breast with its chestnut body and wings makes the brahminy kite (*lang merah*) (*Haliastur indus*) a very distinctive bird. Its soaring flight is very graceful and effortless. The brahminy kite swoops down to capture its prey of fish, crabs, snakes and other reptiles, and even chickens. It is seen in mangroves, open wooded country, near reservoirs, and also close to human habitation. Nests are made in mangrve trees.

A little bigger than a dove, the black-shouldered kite (*lang bahu hitam*) (*Elanus caeruleus*) is a small bird with spectacular large, red eyes. It is mostly white with grey back and wings and black shoulders; its wings are long, but the tail is short. The black-shouldered kite is found in open country, paddy fields and oil palm plantations, feeding on insects, field mice, snakes and grasshoppers. It often hovers before swooping down to catch its prey. Though the black-shouldered kite is widely distributed, it may be threatened by habitat destruction and by the use of chemical pesticides.

The crested serpent-eagle (*lang berjambul*) (*Spilornis cheela*) is a dark brown bird with a black and white crest, and a broad white band across the wings and tail, which is clearly seen when the bird is in flight. The eyes, the unfeathered legs and a conspicuous patch at the base of the long, deeply hooked beak are yellow. The crested serpent-eagle feeds mainly on snakes, but also small mammals and amphibians. Its range extends from lowland forest through secondary forest, rubber and oil palm plantations to mountain forests.

Among the smallest of the birds of prey, the black-thighed falconet is a black bird with a white breast, rufous belly, black thighs and white ring around the face and ear covers. It lives in the forest, from the lowlands to the mountains, and in adjacent open country, feeding on insects and small birds.

The white-bellied sea-eagle (*lang siput*) (*Haliaeetus leucogaster*) is a large, grey bird with white underparts which lives on the coast, near inland rivers, lakes and reservoirs and on offshore islands. Its prey includes fish, birds, turtles, and flying foxes. The sea-eagle forms permanent pairs and has a permanent territory. Its call is a goose-like honking.

A traditional Malay name for the white-bellied sea-eagle was *burung hamba siput* ('slave of the shellfish'). Its role was to scream to warn the shellfish on the shore of changes in the tide so that they knew when to take shelter or when it was safe to emerge to hunt for food.

Found in the mountains, Blyth's hawk-eagle (*lang hantu*) (*Spizaetus alboniger*) is a black and white bird which is found only in Malaysia and Sumatra. It hunts birds and small mammals at canopy level.

The changeable hawk-eagle (*lang hindik*) (*Spizaetus cirrhatus*), an inland bird, is so named because of its two colour phases—dark, when the entire plumage is dusky brown, and a lighter phase when the plumage is variable. It feeds on prey caught at the edge of the forest.

Migrant birds of prey

A large bird with highly variable plumage, the crested honey buzzard (*lang lebah/madu*) (*Pernis ptilorhyncus*) has three phases—dark, intermediate barred, and light. It breeds in Siberia, and flies through Malaysia to the Sunda Islands during September–December, returning to Malaysia in February–April on its way back to Siberia. As many as 500 birds have been sighted in a flock. The crested honey buzzard feeds on honey and bee larvae, mice, insects and reptiles.

The black baza (*lang baza berjambul*) (*Aviceda leuphotes*) follows the same route as the crested honey buzzard, also travelling in large flocks, but some winter in Malaysia. It feeds on large insects and small vertebrates, including small birds.

A rare winter visitor, the black kite (*lang gelap*) (*Milvus migrans*) is usually observed singly in oil palm plantations and peatswamp forests. It feeds on fish, small mammals, lizards and small birds.

The osprey (*lang tiram*) (*Pandion haliaetus*) is found throughout the world, and is closely connected to an aquatic environment, including inland rivers and mining ponds, feeding exclusively on fish. Nests are made on cliffs or trees.

Another visitor to Malaysia during the northern hemisphere winter is the marsh harrier (*lang barat*) (*Circus aeruginosus*), a chocolate brown bird which breeds in western and central Palearctic and western China. Its prey consists of small mammals, birds, amphibians, fish and large invertebrates.

Singalang Burung

The brahminy kite (*lang merah*) (*Haliastur indus*) has special signifcance to the Iban as Singalang Burung, bird-god of war, their most powerful deity. In times of war, Singalang Burung is assisted by the *kenyalang*, a guided weapon based on the rhinoceros hornbill (*Buceros rhinoceros*) which kills the spirits of the enemy, thus ensuring victory in the actual battle.

Seven augury birds act as messengers of Singalang Burung, sending omens to the Iban to provide guidance. To be sent an omen is to have been noticed by the gods, and so great signifcance is placed upon such omens, which must be interpreted correctly.

Though often considered a pest because of its habit of stealing chickens, the brahminy kite is regarded very differently when it appears as the bringer of omens of war. When to fight or not fight, and other major decisions such as the building of a new longhouse depend upon signs from the Singalang Burung. For example, if a brahminy kite lands on a longhouse under construction, the building must be abandoned and a new site chosen.

Nocturnal birds

Malaysia's nocturnal birds can be divided into two groups: owls belonging to the order Strigiformes and nightjars and frogmouths of the order Caprimulgiformes. It is the nightjars which are the most frequently seen because of their habit of resting (and nesting) on the ground in quite open areas in the daytime, rather than high in forest trees as do the owls and frogmouths.

Scops owls

Scops-owls are the tiniest of the owls. Malaysian species include the collared scops-owl (*Otus lempiji*) (top) and the reddish scops-owl (*Otus rufescens*) (middle). It is not uncommon for these owls to have two colour phases, as seen in the two common scops-owls (bottom).

The buffy fish-owl (*Ketupa ketupu*), a large bird with penetrating yellow eyes, derives its name from its fondness for eating fish.

Owls

Though owls are birds of prey, they belong to an order unrelated to other birds of prey such as eagles and kites. Owls are distinguished by their large head with its feathered facial disc and large, forward-directed eyes, which make them look almost human. This is accentuated in some of the true owls, belonging to the family Strigidae, by a tuft of feathers on each side of the head, rather like ears or horns. Barn owls, belonging to the family Tytonidae, have a heart-shaped or oval facial disc, and have longer legs than the true owls.

Their excellent sight and sharp hearing enable owls to hunt at night; they fly silently, helped by their soft plumage with fluffy edges. However, owls give distinctive howls or whistling sounds at night, in contrast to their daytime silence, when they roost in thick cover. Muted colouring helps them to blend with their surroundings during the day or while nesting. Round, white, unmarked eggs are laid in tree cavities left by other birds. Male and female owls are similar in appearance, though female barn owls are larger than the males.

In many parts of the world, owls are regarded with suspicion and fear, and often occur in legends and folklore. The Malay word for owl, *burung hantu* ('ghost bird'), confirms the same place in Malay beliefs. One traditional belief was that the hoot of an owl predicted a death.

True owls

The collared scops-owl (*burung hantu reban*) (*Otus lempiji*) is a tiny creature weighing 65 grams or less.

Like the other owls, it becomes active only at night and roosts quietly on tree branches, sometimes against a tree trunk, in open country during the day. The collared scops-owl eats insects, small mammals and vertebrates (such as lizards and frogs) and arthropods that move about after dark. Its call, a single 'hooup' repeated every 12 seconds, begins soon after dusk. A clutch of 2–5 white, unmarked eggs is laid in a nest in a hollow tree or in the nest of another bird. Little is known about the reddish scops-owl (*hantu merah*) (*Otus rufescens*), a small owl with conspicuous ear tufts which lives in the understorey of lowland forest.

Among the larger owls in Malaysia is the buffy fish-owl (*burung hantu kuning*) (*Ketupa ketupu*), also called the Malay fish-owl, a bird with large, yellow,

The Malay eagle-owl (*Bubo sumatranus*), a large bird with distinctive ear tufts, flies out at dusk to hunt prey. Juveniles are much paler than mature birds.

penetrating eyes. Being strictly nocturnal, it is very difficult to detect during the day unless one knows exactly where it roosts. It feeds on frogs, fish and other small animals and is found in forests near rivers, secondary vegetation and agricultural areas.

The largest of the Malaysian owls is the brown wood-owl (*burung hantu punggor*) (*Strix leptogrammica*), an uncommon resident found in both lowland and montane forests. Adult birds usually live in pairs and remain in well-defined territory throughout the year. They usually have a favoured spot for roosting and return to it day after day. Their nests are located in the main tree trunk, made hollow by decayed branches, where the female lays two white eggs, incubating them by herself, being fed by her mate at intervals during the night. Many insect species, including beetles, crickets and grasshoppers,

The brown wood-owl (*Strix leptogrammica*) is the largest owl in Malaysia.

as well as rodents and small birds, make up their diet.

The Malay eagle-owl (*hantu bubu*) (*Bubo sumatranus*), one of the smallest eagle owls of the world, is found in logged and unlogged forest, mostly in the lowlands. It is fond of bathing in streams or pools of water. At dusk it flies very swiftly from its roosting site, calling a loud and deep 'whoo' or 'hooa-who', ending in a deep groan, but it also emits other calls associated with old folklore.

Barn owls

A pale, ghostlike, medium-sized owl with a heart-shaped facial disc, the barn owl (*jampuk kubur*) (*Tyto alba*) is found throughout the world and is the most widely known and studied of all the owls. Its unusual calls and shrieks, silent flight and gleaming white appearance in the dark, as well as its connection with old, ruined buildings, give it a close association with ghost stories. The barn owl is usually seen at dusk; in the glare of headlamps, especially along quiet village roads, it appears gleaming white. It prefers open habitats, including coastal plains and agricultural lands, human settlements and old buildings. Though nocturnal, it can occasionally be seen during the day, alone or in pairs. Its diet varies greatly depending on habitat, but includes shrews, mice, small birds, frogs, bats and fish. In Peninsular Malaysia, this owl has been used for biological control of rats and snakes. Roosts are provided in oil palm plantations and paddy fields to encourage the barn owl to breed in these areas.

Very poorly known, the bay-owl (*jampuk pantai*) (*Phodilus badius*) is a shy forest owl of similar size and colouring to the barn owl, but with an oval facial disc, which inhabits thick forests up to 1500 metres. Hunting in the dark of night, its prey includes large insects, frogs, lizards, small birds and mammals. During the day, it roosts in nests made in cavities in tree trunks.

The bay owl (*Phodilus badius*) is a shy forest species, roosting during the day in tree holes.

Nightjars

The nightjars, another family of nocturnal birds, are entirely insectivorous. Short-legged birds, they have a net of bristles around the beak, used for catching insects while flying. They roost on the ground during the day and fly out in search of their prey at dusk, in a slow, flapping manner.

The long-tailed nightjar (*burung malas*) (*Caprimulgus macrurus*) emits a rich, deep 'tchoink' at a steady rate of about 3-second intervals, followed by a low growl. This nightjar is quite common along deserted roads in villages and near forest edges, and also in wooded areas, including mangroves, up to about 1200 metres. It usually calls a few times at dawn or dusk while flying or from its perch. During the day, the nightjar can be seen on the ground under trees, well camouflaged by the dead leaves, only flying a short distance away when a person comes very close. No nest is made. Instead, two

Burung malas ('lazy bird') seems an appropriate name for the long-tailed nightjar (*Caprimulgus macrurus*) when it is seen sitting on the ground during the day, moving only if disturbed.

mottled, buff eggs are laid on the ground, and it is there that the mother also cares for her young.

A large bird, with dark brown-black barred plumage and prominent ear tufts, the Malaysian eared nightjar (*taptibau*) (*Eurostopodus temminckii*) becomes active at dusk and dawn, emitting a loud 'tap-ti-bau' call. The Malay name for this bird is derived from its call. A common bird of forest edges and heath forest below 1200 metres, it usually prefers open scrub near forest.

Frogmouths

The frogmouths, related to nightjars but adapted to life in the forest, have a very wide mouth for the capture of insects on the forest floor and from branches. They have mottled, camouflaged plumage and, unlike the nightjars, roost upright on tree branches during the day. Nests are made in trees, usually in the fork of a branch, and one or two white eggs are incubated by both male and female birds.

The most frequently encountered of all the four frogmouth species of Malaysia, Gould's frogmouth (*segan bintik mas*) (*Batrachostomus stellatus*) is a medium-sized reddish brown bird, with a broad bill lined with bristles, which prefers lowland rainforest.

The large frogmouth (*segan besar*) (*B. auritus*) is an uncommon species of the lowland rainforest. It is rarely seen, staying motionless on the branches in the canopy during the day, but is sometimes found in low bushes along streams.

A Jahut carving of the nightjar spirit which lives in the forest and comes to the village at night. Children are not allowed to imitate the 'pok pok' sound of the nightjar as doing this will result in the spirit entering the house. The children would cry and not be able to sleep.

Gould's frogmouth (*Batrachostomus stellatus*) is the most frequently seen of Malaysia's four frogmouth species. Unlike other nocturnal birds, it makes its nest on a tree branch.

Nocturnal bird lore

Walter W. Skeat included in his *Malay Magic*, published in 1900, a collection of Malay bird folklore, including a number of sayings and beliefs relating to nocturnal birds.

Seperti tetegok di-rumah tinggal
Like the nightjar in a deserted house

The nightjar is a solitary bird with a monotonous call. This saying signifies the solitude and loneliness of a stranger in a Malay kampong.

Walter W. Skeat
MALAY MAGIC
With an Introduction by Hood Salleh

Owls were often portrayed as the bringers of death. However, other beliefs also existed. The buffy fish owl (*Ketupa ketupu*) was thought to derive its alternative Malay name, *ketumbuk ketampi*, from its habit of stepping on the edges of its own wings while fluttering the upper section. The resulting sound resembles that of winnowing (*tampi*). A smaller owl, the *jampuk* (*tampi*) (perhaps *Otus scops*) was believed to enter the hen-house and live on the intestines of fowls, using a charm to extract them painlessly.

Yet another legend about the nightjar is that it is a woman who, while husking paddy by moonlight, was turned into a bird because of a quarrel with her mother.

Perhaps also referring to a nightjar is the belief that the *baberek* (or *birik-birik*) bird is the forerunner of misfortune as it was believed to fly in the train of the spectre huntsman (*hantu pemburu*) who roamed the forest with several ghostly dogs; his appearance was the sign of the coming of disease or death. Many charms were used to ward off the influence of the spectre huntsman.

Songbirds

Birds of the order Passeriformes (all the perching birds), which contains more than half the world's birds, are sometimes called songbirds because of their ability to sing very well, which they do mostly during courting or nesting. Malaysia has a wide range of these birds, which can be found in all types of habitat, from rainforest to urban gardens. Some species have traditionally been kept as cage birds because of their melodious songs.

A yellow-vented bulbul (*Pycnonotus goiavier*) feeding its young. Nests are often made in urban gardens, even in quite small bushes.

How a bird sings

Voice box

Birds have a complex respiratory system. The lungs (1) are connected to air sacs (2) filled and emptied by the chest muscles. To sing, the bird closes a valve in one of the two bronchi (3) between the lungs and voice box (4), allowing the air to compress in the sacs. Air pressure in the clavicular sac (5) around the voice box (6) forces very fine membranes (7) into the bronchial passage (8), closing it momentarily. Voice box muscles (9) then become tense (10, 11), pulling the membrane back to reopen the bronchial passage. Air rushing across the tensed membrane vibrates it in song; the pitch rises with increase in tension. The independent working of each pair of muscles in the voice box makes it possible for a bird to sing different notes.

Murai

The magpie robin (*murai kampung*) (*Copsychus saularis*) is a good singer with loud, varied and melodious songs, which sometimes imitate calls of other birds. It is usually one of the earliest songsters to be heard in the morning, often well before dawn. The glossy blue-black plumage of the male is replaced by dull grey in the female. While often seen perched on trees with its tail cocked, the magpie robin usually descends to the ground to feed. It makes an untidy cup nest of dried leaves in holes, often in buildings; a clutch contains 3–5 eggs, pale blue-green, blotched and mottled with reddish brown. A very common bird of gardens, towns, human habitations, agricultural lands and recreational areas, it is thus not protected under the Protection of Wildlife Act 1972 and trading of this species is common.

Like its close relative the magpie robin, the white-rumped shama (*murai batu*) (*Copsychus malabaricus*) is also a good songster. Very popular as a cage bird, especially among the Chinese, it is one of the main species in bird-singing contests. It is often trapped in its habitat, the lower canopy of lowland forest, for trade. A restricted number of licences are issued each year for capture of the white-rumped shama. Usually found singly (sometimes in pairs), it constantly flits about in the understorey of the lowland forest where it builds its nest in tree holes, using twigs, dried leaves and grasses.

Mynas

The common myna (*tiong gembala kerbau*) (*Acridotheres tristis*) is one of the most common birds of the country with a wide distribution, and thus is not protected under the Protection of Wildlife Act. A very highly adaptable bird, occurring everywhere, even on mountains where there are human settlements, it is very bold and can become tame in the presence of humans. Some can be trained to imitate human speech and other sounds made by animals and machines. Common mynas form long-lasting pair bonds, are quite gregarious and usually forage in big groups. In towns, they sometimes roost in large numbers alongside other birds such as swallows, sparrows or crows. Usually omnivorous, they feed on berries, fruits, insects and kitchen refuse. They can be very aggressive and quarrelsome with other bird species or small mammals and reptiles. Normally they nest in holes or cavities, often in buildings or under roofs. Both sexes, which are similar in appearance, take part in building the untidy nest from twigs, dry grass and straw. Eggs (3–5 in a clutch) are a glossy, unmarked, bluish white.

The straw-crowned bulbul (*Pycnonotus zeylanicus*), the largest of the bulbuls, usually sings in duets.

The jungle myna (*tiong hutan*) (*Acridotheres fuscus*) is slightly smaller than the common myna and slightly darker in plumage with a conspicuous tuft of feathers over the base of the bill and forecrown. It lacks yellow markings on the face, unlike the common and hill mynas, but has orange-yellow legs, feet and bill. As widely spread as the common myna (and apparently an ecological competitor), but not as common, it is not protected under the law. A highly gregarious bird of forest edges and cultivated areas, it forms long-lasting pair bonds. Its song is typical of mynas, consisting of a sudden chorus of repeated whistling phrases that are rather discordant and out of tune.

The hill myna (*tiong mas*) (*Gracula religiosa*) is a popular cage bird with a good ability to

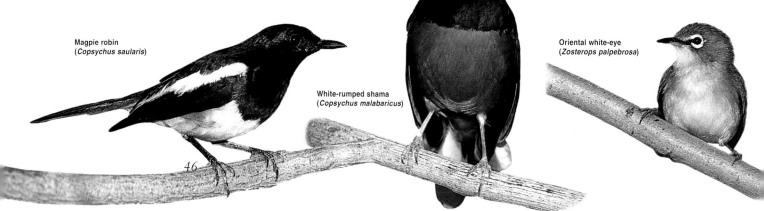

Magpie robin
(*Copsychus saularis*)

White-rumped shama
(*Copsychus malabaricus*)

Oriental white-eye
(*Zosterops palpebrosa*)

Its bright colouring and distinctive call make the black-naped oriole (*Oriolus chinensis*) perhaps the most easily identified of the garden birds.

mimic human voices, other bird calls or any other sound, such as machines and automobiles, if these are heard frequently enough. Living in the canopy of lowland rainforest, it is often heard giving a very clear 'tiong' whistle while perching high on tree branches or flying in small groups. In flight, the clear white patch on the wings and its fast wing beats are very conspicuous.

Bulbuls

As the most common of all the bulbuls, the yellow-vented bulbul (*merbah kapur*) (*Pycnonotus goiavier*) is very widely distributed throughout Malaysia and is not a protected species. It is very easily observed at close quarters and is rarely silent; its continual chattering in shrubs and gardens around us is always cheerful. Like the magpie robin, it is an early riser and provides wake-up calls at daybreak. It feeds mostly on fruits, berries and many kinds of insects.

The red-whiskered bulbul (*merbah telinga merah*) (*Pycnonotus jocosus*) is only common in the northern states, especially Perlis, Kedah, Kelantan and Perak, usually in gardens and parks, villages, scrub and secondary growth. It is a very popular cage bird and any which escape can easily establish a wild population. A licence is necessary to keep the red-whiskered bulbul as a cage bird since it is protected under the Protection of Wildlife Act.

Usually performed in duet, bubbling, melodious songs of the straw-headed bulbul (*barau barau*) (*Pycnonotus zeylanicus*), the largest of the bulbuls, can be heard over a long distance. It is also protected under the Protection of Wildlife Act, but a licence can be obtained by those wishing to keep it as a cage bird. This bulbul lives in scrub and woodlands, especially near water, and can often be heard along river banks.

Songbird contests

The traditional keeping of cage birds solely for companionship has evolved in recent decades into the rearing and training of birds for the sport of formal bird-singing contests. Contests are held early in the morning, with birds hung a fixed distance from each other. Cages are kept covered until the contest begins, when all owners must leave the arena. The birds are judged on physique and display in addition to their singing, though it is the song which is allotted most points. In the judging of songs of birds such as the white-rumped shama and oriental white-eye, points are allotted for characters such as loudness, variety and stamina. Judging of doves is, however, done differently as their 'song' is unlike that of the other birds. Marks are given for each section of the song—opening, middle and final sound—as well as for the melody and the quality and clarity of the overall sound.

Each contest has a number of rounds, each with a different judge, and the points are totalled at the end of a contest. Such contests become very competitive because the value of a bird increases with the number of contests it has won.

Originally, all the songbirds were captured from the wild. However, with declining wild populations and restrictions on the capture of some species, captive breeding of such birds is becoming a more important source.

Oriental white-eye

Another popular cage bird, the oriental white-eye (*mata putih*) (*Zosterops palpebrosa*) is often seen in small groups, moving through vegetation in search of insects. It is a small, lively bird, fairly common in mangrove forests and coastal gardens and plantations.

Babblers

A small, dull brown babbler with a heavy bill and short tail, Abbott's babbler (*rimba riang*) (*Trichastoma abbotti*) is probably the most common of all the babblers in the country. It inhabits secondary forests, shrubs and forest edges where its loud, distinctive whistle is often heard very early in the morning. It is usually found singly skulking close to the ground.

Black-naped oriole

The black-naped oriole (*dendang selayang*) (*Oriolus chinensis*) is another bird which has moved very successfully from coastal to urban areas. A bird with a very melodious whistle, it feeds on the fruit of garden trees as well as large insects. Like many other birds, it is very active in the morning and evening.

Bird owners watch anxiously as their birds are judged in this bird-singing contest in central Kuala Lumpur.

In preparation for bird-singing contests, birds are taken in the early morning to open areas such as this equipped with tall poles. where they can 'practise' by listening to the songs of other birds.
INSET: An example of a traditional bird cage with Siamese influence.

mon myna
dotheres tristis)

Jungle myna
(*Acridotheres fuscus*)

Hill myna
(*Gracula religiosa*)

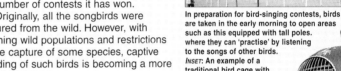

47

Swifts and swiftlets

Swiftlets are best known to Malaysians as the birds whose edible nests are the source of birds' nest soup and other gourmet delicacies which utilize this seemingly unlikely source of food. However, it is only two species, the black-nest swiftlet (Aerodramus maximus) and the white-nest swiftlet (A. fuciphagus), which make these commercially valuable nests, though Malaysia has 13 species of swifts and swiftlets.

House swifts (*Apus* sp.) nesting under the roof of an old house, a practice which makes them unpopular as their nests are dirty and are a fire hazard.

Swifts and swallows

'Swift' is the name given to the family of birds which fly with great speed. The true swifts, unlike swallows, rest by clinging onto cave walls and buildings, and glide with extended wings. Swallows and tree swifts, which belong to a different family, perch on trees or overhead wires and fly with partly closed wings. The smaller swifts, known as swiftlets, fly with less speed, often changing direction and tilting from side to side.

Typical swifts

All swifts belong to the family Apodidae and have a compact body, long wings, short neck and very short legs with sharp, pointed claws. A reversible hind toe on their feet enables them to cling onto steep rocks. Swifts can fly over 20 kilometres in a day, eating, drinking and bathing while flying.

Swifts feed entirely on swarming insects trapped and picked from under the wings while in flight. Most nest in large colonies in a tree hole or on a cave wall, using plant materials held together by saliva. Hatchlings are fed by both parents on insects regurgitated from their throats. Swifts are of little economic importance except for their consumption of large quantities of harmful insects.

The house swift of the genus *Apus* builds nests of mud and grass in buildings, creating a fire hazard as these nests easily catch fire from a wisp of cigarette smoke. Moreover, the droppings from the nests are dirty and tend to deface buildings.

Cave swiftlets

Malaysia has five species of swiftlets belonging to three genera: *Aerodramus*, *Collocalia* and *Hydrochous*. All nest in rock shelters, caves, buildings or culverts. The giant swiftlet (*Hydrochous gigas*) makes shallow, mossy nests on mountain cliffs while the white-bellied or esculent swiftlet (*Collocalia esculenta*), the smallest swiftlet in Malaysia, nests in cave mouths, underneath bridges and in roof spaces of houses, using lichens, mosses and ferns.

Unlike the giant swiftlet and the white-bellied swiftlet, all *Aerodramus* swiftlets can fly in total darkness using echolocation. Of no commercial value, the mossy-nest swiftlet (*A. vanikorensis*) builds nests of mosses, liverworts and ferns with little nest cement. The black-nest swiftlet (*A. maximus*) nests in limestone and sandstone caves where it constructs the typical black nests of saliva cement mixed with many feathers. However, the best quality edible nests are made by the white-nest swiftlet (*A. fuciphagus*) found on offshore islands, in old houses or in limestone caves. The nest cup is made entirely of firm saliva cement mixed with a few feathers. Only a row of six or seven small feathers on the outside, and another four feathers on the inner side of the tarsus of the black-nest swiftlet differentiate it from the other commercial species.

White-bellied swiftlets (*Collocalia esculenta*), Malaysia's smallest swiftlets, make nests of moss and lichens in houses as well as in cave mouths.

A mossy-nest swiftlet (*Aerodramus vanikorensis*) in its nest of moss. Like the edible nest swiftlets of the same genus, this swiftlet uses echolocation to fly in total darkness.

Swiftlet biology

The black-nest swiftlet lays one egg, while the white-nest swiftlet lays two in a clutch during the breeding season (September–April). The swiftlets need more than a month to build their nests and another to incubate the eggs, as well as a further two months for the young to fledge.

The edible-nest swiftlet of the genus *Aerodramus* showing its short bill and compact body.

Nests are constructed by the regurgitation of saliva from a pair of lobed salivary glands under the tongue. As a pad of hardened saliva cement adheres to the rock wall, the nest takes on a crescent shape. Subsequently, the bird smears the edge of the nest with saliva from the sides of its bill, laying down layers of nest cement until a small, cup-shaped nest is formed. Sometimes, the bird plucks its own wing feathers to cushion the nest.

Conservation and management

Birds' nests are protected in Sarawak under the Wildlife Ordinance while in Sabah the control and collection of birds' nests is regulated under the Birds' Nests Ordinance. However, the swiftlet population has declined in recent years due to overharvesting by illegal poachers. Birds' nest production can be a sustainable activity if nest harvesting is adapted to the swiftlets' breeding cycles and the birds are given sufficient time to breed and raise their young. The long life expectancy of the swiftlets (15–20 years) has sustained the present production of birds' nests.

DISTINGUISHING FEATURES OF MALAYSIAN SWIFTLETS

SPECIES	SCIENTIFIC NAME	WING SPAN (mm)	TAIL LENGTH (mm)	WEIGHT (g)
Black-nest swiftlet	*Aerodramus maximus*	125–135	48–52	15–20
Giant swiftlet	*Hydrochous gigas*	140–160	55–65	35–40
Mossy-nest swiftlet	*Aerodramus vanikorensis*	115–125	45–50	10–15
White-bellied swiftlet	*Collocalia esculenta*	95–105	35–45	10
White-nest swiftlet	*Aerodramus fuciphagus*	115–125	42–48	10–15

Collection and processing of birds' nests

Location of edible nest swiftlets

South China Sea

BRUNEI

Sabah
Gomantong Caves ●
Kuamut Caves ●
Madai Caves ●
● Niah Caves

Sarawak

Kalimantan

N

Collecting swiftlet nests in Gomantong Caves, Sabah, a major source of both black and white nests.

Black-nest swiftlets (*Aerodramus maximus*), one of the two species of edible birds' nest swiftlets in Gomantong Caves, Sabah.

Collecting edible birds' nests

The trade with China in birds' nests has probably been in existence for about 600 years, originally as barter trade. Birds' nests, which are collected before the breeding season of the swiftlets, are found on the ceiling of cave taverns as high as 50 metres from the ground in the Niah Caves of Sarawak and the Gomantong, Madai, Kuamut and other small caves in Sabah. The Niah Caves have black-nest swiftlets. In Gomantong, white-nest swiftlets are found in Simud Putih Cave, while black-nest swiftlets occupy Simud Hitam Cave. The Madai Caves contain both species of edible nest swiftlets.

The nest harvesting technique used depends on the height of the cave and the substrate below. In some caves in Sabah, rattan ladders are used to reach the ceiling. In Sarawak, the collectors normally climb up a vertical pole 20–50 metres high to reach a collecting platform at the top. In caves with underground streams, a fishing net is laid out to collect the nests.

In the Niah Caves of Sarawak, the collecting tools consist of a set of 6–8 lengths of socketed bamboo stems each 3 metres long. The top length has a hoe-shaped iron head for scraping the nests, and also a lighted candle. The collectors are highly skilled in climbing up the poles to reach the nests. As the nests are scraped off the cave wall, another collector picks up the fallen nests from the cave floor.

A traditional thanksgiving ceremony is usually held at the cave entrance to appease the spirits of the mountains before each harvest season. After the ceremony, the cave is normally closed to visitors.

Birds' nest processing

The quality of edible birds' nests varies according to the location of the nest caves. Some black nests contain red basal cement, while others are white. The red colouring is due to the presence of iron compounds in the rock to which the nests are attached. White nests from some caves can turn red on exposure to moisture, while white nests built in houses are usually yellowish, depending on the nature of the wood on which they are built.

A kilogram of unprocessed black nest with few black feathers may fetch RM1,000 before processing, while the price of white nests can be five times higher. White nests are usually very clean with few strands of feathers or eggshell fragments. They can be easily cleaned before cooking by soaking in water for about 30 minutes. However, if tap water containing chlorine is used to clean birds' nests, the nests will not expand and the feathers cannot be removed with ease.

Nests are scraped off cave walls with a hoe-shaped tool.

Manual picking of the feathers from the soaked black nest.
INSET: Top-quality unprocessed white nest of the edible-nest swiftlet.

The best quality black nest after processing is called *taohu yen*, after the town of Tawau in Sabah where birds' nests are traded.

The red phase of the black nest after processing is called *siah yen*; the colour was believed to have come from the blood of the swiftlets.

Black nests are processed by first soaking in water for 12–48 hours, depending on the dryness of the nests. The large feathers are then separated from the nest cement using a pair of fine forceps. Fine feathers are removed by a flotation technique: the nests are swirled in a circular container with a few drops of vegetable oil. Long and meaty strands of nest cement are rearranged in crescent-shaped moulds to produce premium grade cleaned nests, the colour of which should be the shiny yellowish white of the raw nests. Chemically processed nests are generally a faded yellowish colour. To be sure of quality, it is better to purchase nests with black feathers embedded in them. Colour and smell can be used as a guide to distinguish between genuine and fake nests. Genuine nests do not turn mouldy when kept in an airtight container.

Birds' nest soup

Birds' nests soup cooked with Chinese herbal ingredients has traditionally been used for treating maladies of the lungs, as a food to rejuvenate the skin, and as a food tonic. It is considered a cooling food by the Chinese.

There are many ways of preparing birds' nest soup. The most common way is to double boil the nest materials with rock sugar or chicken stock in a covered porcelain pot for two hours. Other methods of preparation include cooking with ginseng, red dates, honey or dried longan. The cooked nest is best consumed while still warm at night and before sleeping, or in the early hours of the morning.

Nutritional value of birds' nests

Birds' nests contain a water soluble glycoprotein with the properties of both protein and carbohydrate. Being water soluble, the amount of glycoprotein lost during soaking depends on the length of time the raw nest is soaked. The percentage of glycoprotein and its health-giving functions have not been clearly established.

COMPOSITION OF CLEANED, DRY BIRD'S NEST

Protein 85%	**Ash** 2.5%
Fat 0.3%	**Water** 12.0%

Trace minerals	**Amino acids**
phosphorus	amide
calcium	humin
iron	arginine
arsenate	cystine
sulphur	histidine
	lysine

Energy value 345 kcal/100 g

Pheasants and the jungle fowl

Pheasants and the jungle fowl, together with other game birds, belong to the order Galliformes. Members of this order differ from most other birds in spending most of their time on the ground rather than perching in trees. While Malaysia's eight pheasant species (which include three endemic species) are rarely seen birds as they live deep in the forest, the red jungle fowl is frequently sighted near forested areas.

A male great argus pheasant (*Argusianus argus*), a rarely seen bird of the rainforest, whose beautiful long tail feathers can give him a total length of up to 2 metres.

Pheasants

Malaysia's pheasants, all magnificent birds with many 'eyes' (ocelli) on their plumage, are found in rainforest, from lowlands to mountains, depending on the species. They dwell in the ground storey, seeking food (insects, berries, fallen fruits) solely on the forest floor, where they also make their nests under bushes or fallen trees. It is usually the female alone who incubates the eggs and looks after the young. At night, pheasants roost on tree branches above the forest floor.

Most male pheasants have their own dancing grounds where they display and call to attract females. Each tries to find the best location for his dancing ground, a level area either on flat land or on a mountain ridge. The bird clears the area of any vegetation and keeps it free of litter. These grounds are guarded against intruders. Displays by different species vary, but all are designed to utilize their many tail feather 'eyes' to attract females.

Firebacks

The crested fireback (*ayam pegar*) (*Lophura ignita*) takes its name from the erect crest on both males and females. The male has spectacular colouring—an overall plumage of dark bluish violet with a reddish orange patch on the back merging into maroon on the rump and white central tail feathers—while the female is brown with white and black scale-like feathers on the throat, breast and belly. Both sexes have red legs. Crested firebacks inhabit the forest floor, in groups of 5–6, in lowland rainforest, generally near river banks. Insects and fruits are found in the leaf litter. They are often seen on riverside trails in Taman Negara. Males whir

A male crested fireback pheasant (*Lophura ignita*) (above) and a male crestless fireback pheasant (*Lophura erythrophthalma*) (below), both of which try to attract females by whirring their wings.

their wings in an effort to attract females.

Both sexes of the crestless fireback pheasant (*merah mata*) (*Lophura erythrophthalma*) have black plumage. The male is metallic bluish black with buffy brown tail feathers and a reddish chestnut rump, while the female is totally black except for scarlet facial skin. Those in Sabah and Sarawak have a deeper colour than birds found in Peninsular Malaysia, probably because they are found at slightly higher altitudes. Crestless firebacks are found in logged and unlogged lowland forest and hill forest up to 800 metres, usually singly, in pairs or in small family groups of 5–6. Like the crested fireback, the male of this species whirs its wings during display or when alarmed.

Two Keningau Murut women in Sabah with the beautiful tail feathers of the great argus pheasant (*Argusianus argus*) adorning their headdresses.

Argus pheasants

The two argus pheasant species occupy similar habitats—rainforest up to 900 metres—and have similar habits. In both species each male has its own dancing ground, a small clearing about 1.5–2 metres square on level ground.

Quails

Among its game birds, Malaysia has two quail species, the blue-breasted quail (*pikau*) (*Coturnix chinensis*) and the barred button quail (*puyuh*) (*Turnix suscitator*). Although similar in appearance, anatomical differences place these two birds in different orders. The barred button quail (below left), which is reared for both its meat and its very popular tiny, speckled eggs, is one of the few species whose female is more colourful than the male.

Examples of beautifully carved quail traps (*jebak puyuh*) are shown in a painting by Shafie bin Haji Hassan, and at right. A decoy bird is placed inside a trap which is activated by one of a variety of mechanisms to capture its victims. Although wood is used for the front of traps, a number of different materials are used for the back portion, including basketware and coconut shells.

At top right is a quail kite. Its name is derived from the bird in the centre of the kite. In villages on the east coast of Peninsular Malaysia, kite flying is a traditional pastime.

Males are polygamous and use loud calls and display their ornamental feathers to attract females. The two sexes are solitary, meeting only during mating seasons. After mating, the female leaves the male to lay her eggs.

As in many birds, the male of the great argus (*kuang raya*) (*Argusianus argus*) is much more attractive than the female, with long tail feathers adorned with small white spots, and large secondary wing feathers decorated with golden eyespots. These features are used by the male to show off when displaying.

The crested argus (*kuang gunung*) (*Rheinardia ocellata*) is very limited in distribution, occurring only in Vietnam and Peninsular Malaysia, where populations are found on only a few mountains in Taman Negara—Gunung Tahan, Gunung Rabung, Gunung Mandi Angin and Gunung Gagau.

Peacock pheasants

The mountain peacock pheasant (*kuang cermin*) (*Polyplectron inopinatum*) is endemic to Peninsular Malaysia, inhabiting montane forests between 915 and 2200 metres. Both males and females have dark brown plumage with small, buffy spots, but females can be distinguished by their smaller body, shorter tail feathers and one spur on each leg (males have two spurs). The distribution of this species is limited to the central Titiwangsa mountain range, extending from Gunung Bintang Hijau in the north to Genting Highlands in the south, and also Gunung Benom in the Krau Wildlife Reserve in Pahang.

In contrast, the Malaysian peacock pheasant (*merak pongsu*) (*Polyplectron malacense*), also endemic to Peninsular Malaysia, is usually found in lowland forest on level ground up to 200 metres high, singly or in pairs. Both sexes are small and heavily feathered, but males have a crest and one spur. During the breeding season, both males and females emit loud clucking sounds. Males make a small dancing ground where they display their beautiful feathers laterally, often offering food like small insects, grubs or fruit to the female between intervals of dancing. Mating usually occurs only with the consent of the female. After mating, the male leaves the female to lay a single egg on dry ground with a little shelter.

A male and female Bulwer's pheasant (*Lophura bulweri*), which is found only in Sabah and Sarawak. As with many other birds, it is the male which is the more colourful of the pair.

Bulwer's pheasant

Endemic to Borneo, Bulwer's pheasant (*burung pakiak*) (*Lophura bulweri*) is found in both Sabah and Sarawak in hill forest up to 1000 metres high, singly or in small groups. Its shy, quiet behaviour makes it very difficult to see. The male has a very impressive

display during which its wattle (loose skin hanging from the throat) distends to form a blue ribbon against the background of its glossy white tail. Its overall body plumage is dark metallic blue with a maroon breast, but the female is dull brown with bluish facial skin; both have reddish legs.

Peafowl

Believed to be extinct in Peninsular Malaysia, the Javanese green peafowl (*merak*) (*Pavo muticus*), similar in appearance to the related Indian peafowl (*Pavo cristatus*), but with green colouring, used to inhabit forest edges on the east coast and in Perak. It was the largest in the pheasant group, with the longest tail feathers. However, many are kept in captivity throughout the world and the Department of Wildlife and National Parks is planning to reintroduce this species to its natural habitat. Like the great argus, the male displays its beautiful feathers elaborately in order to attract females.

The red jungle fowl

The red jungle fowl (*ayam hutan*) (*Gallus gallus*) is believed to be the ancestor of domestic chickens, but no one really knows from which species of jungle fowl the domestic chicken originated. The jungle fowl can be crossbred with domestic chickens, resulting in a bird which is smaller than the domestic chicken but bigger than the jungle fowl and wilder than the domestic chicken. The male jungle fowl resembles a male domestic chicken but has a slimmer body, grey legs and typically large white ear patch, while the female is brownish with dark brown streaks. Both are very wild and easily alarmed.

The red jungle fowl is found in secondary forests, oil palm and rubber plantations and forest edges, and nests on the ground. It moves around in groups of 5–8 birds, mostly females and juveniles. The call of the jungle fowl, though similar to that of the domestic cockerel, can be distinguished by the shorter last note.

A male Malaysian peacock pheasant (*Polyplectron malacense*) with beautiful green 'eyes' on its wing and tail feathers. INSET: An egg of this species.

The Javanese green peafowl (*Pavo muticus*), a magnificent bird which was hunted so extensively it has become extinct in Malaysia. However, it may be reintroduced to its natural environment in the future.

The red jungle fowl (*Gallus gallus*), which is believed to be the ancestor of domestic chickens.

Jungle fowl motifs

Of great cultural significance to the Malays, the red jungle fowl is often incorporated in the design of traditional handicrafts. Examples shown here are two pillow ends (the top one of gilded silver, the lower one of gold thread embroidery) and the detail of a length of *songket* (traditional sarong of silk interwoven with gold thread worn on ceremonial occasions) with the red jungle fowl in its design.

Pigeons and doves

Of the world's 290 pigeon and dove species which form the family Columbidae, Malaysia has only 26 species, all residents, divided into three groups: green pigeons (12 species of small, brightly coloured arboreal birds), ground doves (10 species of highly iridescent birds usually found on the ground) and imperial pigeons (4 species of large, arboreal birds with a metallic sheen).

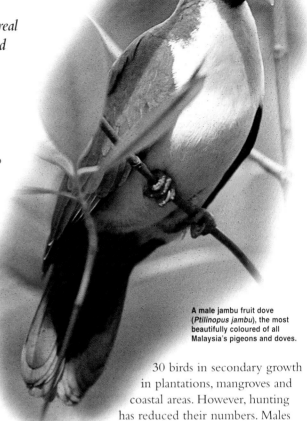

A male jambu fruit dove (*Ptilinopus jambu*), the most beautifully coloured of all Malaysia's pigeons and doves.

Basket weaving

In Sarawak, many basketwork designs are taken from the animals and plants which play a role in the life of the weaver. The green pigeon's eye design is shown above. This bird is thought to ensure success in any venture.

Characteristics

There is no scientific difference between pigeons and doves, but the term 'dove' tends to be used to refer to species with pointed wings and rather long tails. Pigeons and doves are mostly plump-bodied birds with short legs. They have a small head with a short, thick bill. Mostly gregarious birds, they feed on fruits, seeds and buds. Their flight is strong and direct. Nests are built on small trees by placing dried twigs on top of each other in an untidy manner to form a rough platform. Their white eggs are usually laid in a clutch of two.

Green pigeons

The thick-billed green pigeon (*punai lengguak*) (*Treron curvirostra*), a very conspicuous bird with a bright yellow-green bill with a red base, red feet and blue-green skin around the eyes, is usually found in large flocks in fruit trees in the lower zone of montane forests. Male birds have maroon wings and back, females green.

Distinguished by its small size, the little green pigeon (*punai siul*) (*Treron olax*) is usually found in pairs or small flocks feeding on figs in submontane forests and parks up to 1100 metres. Male and female birds have different colouring: both have a grey head and green breast, but males have an orange breast band and maroon wings, while females have a pale throat.

Much more common than the little green pigeon is the pink-necked green pigeon (*punai gading*) (*Treron vernans*), which can be seen in flocks of up to 30 birds in secondary growth in plantations, mangroves and coastal areas. However, hunting has reduced their numbers. Males have colourful plumage: a grey head, pinkish purple neck and an orange band on the lower chest. In contrast, females are dull green. This pigeon can be distinguished from other green pigeons by its grey tail with a black band. It also makes a whistling call in contrast to the 'coo' of most other pigeons.

Most brilliantly coloured of the pigeons and doves is the jambu fruit dove (*punai jambu*) (*Ptilinopus jambu*). The male has a crimson face with a black throat patch, green upper parts, and white underparts except for a pink patch on the throat. The female is dull green, but both sexes have a yellow bill, red legs and a pale ring around the eyes. In contrast to its bright colouring, the jambu fruit dove is a quiet, solitary bird, usually seen on fig trees in forests up to 1100 metres.

Ground doves

The bird common throughout the world near human settlements is the rock pigeon (*merpati*) (*Columba livia*), which is usually dependent on people for food. As in other countries, these birds can be seen in large flocks in places where they are routinely fed, such as at temples. There is small-scale breeding of these birds for food in Malaysia.

Endangered because of hunting and the collection of their eggs for food, the largest of the ground doves, the Nicobar pigeon (*punai Nikobar*) (*Caloenas nicobarica*) is a very colourful bird with long, multicoloured neck feathers and an incongruous white tail. It is found on offshore

Colourful pigeons

A green-winged pigeon (*Chalcophaps indica*), often seen feeding on seeds on the ground in oil palm plantations.

A pied imperial pigeon (*Ducula bicolor*), a distinctive cream-coloured bird of offshore islands and coastal areas.

A pink-necked green pigeon (*Treron vernans*), a coastal species which, like the other green pigeons, is strictly arboreal.

A Nicobar pigeon (*Caloenas nicobarica*), a very colourful species found only on offshore islands, feeding on a papaya in the Kuala Lumpur Bird Park.

Cage birds

Spotted-necked dove (*Streptopelia chinensis*)

Peaceful dove (*Geopelia striata*)

As seen in the painting (left) by Shafie bin Haji Hassan of a man emerging from a kampong house with a pet bird, Malays have traditionally kept doves as cage birds, loved for their gentle nature and melodious voice. The two favoured species are the *tekukur* or spotted-necked dove (*Streptopelia chinensis*) and the better known *merbuk* or peaceful dove (*Geopelia striata*), which is also entered in bird-singing contests.

In such contests, the song of the doves is judged differently from other songbirds as the nature of their song is different. Successful birds can be of high monetary value.

Because of their popularity as cage birds, these two doves are heavily trapped, sometimes by a beautifully crafted trap such as that shown at left. A decoy bird is placed inside to attract another bird which, as it nears the front of the trap, steps on a 'trigger' which releases a net to ensnare the bird.

islands on both the west and east coasts of Peninsular Malaysia, but not on the mainland.

In the montane forests down to 500 metres, the most common species is the little cuckoo dove (*tekukur api*) (*Macropygia ruficeps*), a long-tailed, brown pigeon which nests in large colonies in dense, low growth. Its loud calls can be heard in the early morning and the evening.

Very popular as a cage bird, the spotted-necked dove (*tekukur*) (*Streptopelia chinensis*) is often trapped by using a decoy in a small cage. Attracted by the singing bird, the doves are caught in the snares. The spotted-necked dove is commonly found on the ground in open country and cultivated areas in small groups. Named for the half-collar of spots on its neck, this bird is brown above and pinkish below. Both sexes take part in building a nest and in incubating the eggs. Its local name, 'tekukur', reflects its melodious, repeated call.

One of the most popular songbirds, the peaceful (zebra) dove (*merbuk*) (*Geopelia striata*) is also heavily trapped for trade as a cage bird. Birds which are prizewinners may be worth thousands of ringgit. The alternative name of zebra dove can be attributed to its striped plumage. Similar in colour to the spotted-necked dove, the peaceful dove is a much smaller bird with blue cheeks.

The green-winged pigeon (*punai tanah*) (*Chalcophaps indica*), which is also known as the emerald dove, takes its name from its iridescent green wings, which contrast with its pink face. It is common in rural areas, on the ground (singly or in pairs) in plantations and forests up to 1200 metres.

Imperial pigeons

By swallowing large seeds whole, the green imperial pigeon (*pergam besar*) (*Ducula aenea*) acts as a seed disperser. It is a large, striking bird with a bulky, pale grey body, green wings and a dark grey tail which is

found in lowland and hill forests, mangroves, coastal areas and islands in small flocks.

Found on offshore islands as well as in mangrove and coastal areas in much larger flocks than the green imperial pigeon is the pied imperial pigeon (*punai rawa*) (*Ducula bicolor*). It is a distinctive, creamy white bird with black flight feathers and tail. Although it is protected under the law, it is a popular target for hunters for both food and sport.

The mountain imperial pigeon (*pergam gunung*) (*Ducula badia*) is a larger species with a pale grey body which contrasts with its maroon wings and the red of its beak and a ring around the eyes. An arboreal bird, usually found in the high canopy, this pigeon has broad, rounded wings, slow wing beats and a deep, booming call.

Conservation

All the Malaysian pigeon and dove species except the spotted-necked dove and the peaceful dove are protected under the Protection of Wildlife Act. However, hunting of some of the protected species is allowed in September each year. The spotted dove and peaceful dove are both hunted heavily because of their popularity as cage birds.

Images of the dove

The cultural significance of the dove is reflected in its use in handicrafts. A beautiful pair of wedding slippers (above), embroidered by a Straits Chinese craftswoman, features a dove, as does the Malay wedding dowry (*hantaran*) (below) in the form of a bird made entirely of 1-ringgit notes.

Pigeons and doves often feed together in flocks, such as these rock pigeons (*Columba livia*) (left) near the Batu Caves on the outskirts of Kuala Lumpur and peaceful doves (*Geopelia striata*) at the Kuala Lumpur Bird Park.

Hornbills

There are 54 species of hornbills in the world, forming the order Bucerotiformes. They are found in sub-Saharan Africa, and from India east to New Guinea and the Solomon Islands. Malaysia has 10 species of these astonishing, long-tailed birds, which in Sarawak have a special place in customs and ceremonies, as they have in many other societies.

Malaysia's hornbills

The distinctive features of the male and female of each species can be clearly seen in this illustration.

1. Helmeted hornbill
   (*Buceros vigil*)
2. Bushy-crested hornbill
   (*Anorrhinus galeritus*)
3. Rhinoceros hornbill
   (*Buceros rhinoceros*)
4. Wreathed hornbill
   (*Aceros undulatus*)
5. Plain-pouched hornbill
   (*Aceros subruficolis*)
6. Great hornbill
   (*Buceros bicornis*)
7. Black hornbill
   (*Anthracoceros malayanus*)
8. Oriental pied hornbill
   (*Anthracoceros albirostris*)
9. White-crowned hornbill
   (*Aceros comatus*)
10. Wrinkled hornbill
    (*Aceros corrugatus*)

Detail of a historical photograph of Iban warriors in ceremonial dress with hornbill feathers.

Characteristics

Hornbills are large, broad-winged and long-tailed forest birds with black or brown and white plumage which is often stained yellow in older birds. Though not easily seen in the forest, their loud calls are distinctive and can be used for identification. The bushy-crested hornbill (*Anorrhinus galeritus*) yaps, the wrinkled hornbill (*Aceros corrugatus*) barks, while the black hornbill (*Anthracoceros malayanus*) makes a retching sound.

Atop their enormous bills, many species have a projection known as a casque. In only one Malaysian species, the helmeted hornbill (*Buceros vigil*), is the casque solid, and carved in the same way as ivory. In previous times, this 'ivory' was a valuable item in trade with China. Carved casques were traditionally worn as earrings by some ethnic groups in Sarawak.

The rhinoceros hornbill (*Buceros rhinoceros*) is featured in the state crest of Sarawak, which is often called 'Land of the Hornbills'.

Nesting

It is not only the appearance of the hornbills which is distinctive. Their nesting behaviour is unique. Like a number of other birds, the hornbills are all hole-nesters. However, unlike other birds, the female hornbill is sealed into this hole, with only a narrow oval slit left open to allow her mate to feed her and the chicks. The female remains in the nest until the chicks are ready to fly.

Nest-building is usually carried out by the female, with the male bringing food and sealing and lining materials. The number of eggs laid varies, according to species, from one to eight. All hornbills lay oval, rather elongated white eggs. The incubation period is between 21 days for smaller species and 42 days for larger birds. The same pair of birds returns to the same nest every year.

Diet

The hornbills are omnivorous, mostly feeding on fruit and berries, but also eat birds' eggs, insects and other small animals. They often fly long distances in search of fruiting trees. Some restrict their search for food to the forest canopy; others range more widely. The oriental pied hornbill (*Anthracoceros albirostris*) sometimes forages on the ground.

Habit and habitat

Being specialized breeders needing large trees, Malaysian hornbills are primarily forest dwellers. However, they are found at various altitudes, and some are also found in mangrove forests. Most live in pairs or small groups, but flocks can number 20–50 birds. A group of 2,000 wreathed hornbills (*Aceros undulatus*) has been recorded flying to their nesting site in Belum, Perak.

The Iban and hornbills

Although it is not considered sacred, the rhinoceros hornbill has special significance to the Iban people of Sarawak, whose name for this bird is *kenyalang*.

It is often hunted and eaten, and the white and black tail feathers used to decorate helmets and other ceremonial dress. Carved wooden hornbills are used in the *gawai kenyalang* (hornbill ritual) to attack the souls of enemies. These carvings vary from the very simple to the very ornate, and some are as large as 2 metres long and 1 metre high. They are first carried in procession, and are then used in the longhouse to reinforce moral values before being hoisted atop tall poles, and fastened in place, with the icon facing enemy territory. The soul of the kenyalang is believed to kill the soul (*semangat*) of the enemy, so ensuring victory in a future attack.

Iban hornbill effigy. The real hornbill's basic outline is realistically observed: the bird has a disproportionately long, heavy beak, a squat body, and straight tail feathers about as long as the body.

Kingfishers and woodpeckers

Malaysia has 15 species of kingfishers and 26 species of woodpeckers and piculets. Though belonging to different orders, both groups are hole nesters and have feet specially adapted to their way of life. Kingfishers are never very far from water as fish is the main component of their diet and their nesting holes are often in river banks. Woodpeckers, on the other hand, utilize trees for both their food source and nesting sites.

To build a nesting hole in a river bank, the kingfisher flies at the bank and uses its bill to dislodge soil until there is enough room for it to place its feet. The claws and bill are used to excavate a tunnel up to 1 metre long, ending in a breeding chamber.

Kingfishers

Kingfishers form the family Alcedinidae belonging to the order Coraciiformes, to which the hornbills also belong. They are a group of brightly coloured birds with a large head, a large bill, short, weak, black or red legs and a short tail. However, size varies, as does colouring. Many are noisy. Kingfishers are usually found near water, in coastal and freshwater habitats. The majority are carnivorous, eating fish, crabs, frogs, lizards, insects and arachnids, but a few eat fruits and berries. Prey is caught by diving into the water or plunging to the ground from their perch on a tree. As birds which are solitary in habit, kingfishers usually hunt alone.

The three front toes of kingfishers are joined for part of their length, making it difficult for them to walk, but making their feet useful tools for shovelling soil away when tunnelling. Kingfishers do not build nests, but dig holes in river banks or termite mounds, or use holes in trees. Their eggs are usually white and unmarked; most clutches have 3–6 eggs. In most species, both male and female excavate the nest and incubate the eggs.

Not as common as in temperate countries, the common kingfisher (*pekaka cit-cit kecil, raja udang*) (*Alcedo atthis*) is a small bird (16 centimetres) with a blue-green back, rufous chest and abdomen, black bill and bright red legs. It is found in open lowland areas (sometimes as high as 1800 metres) near water and feeds on fish, tadpoles and aquatic insects. Nests 0.5–1 metre long are built in steep river banks.

Similar to the common kingfisher, but with darker colouring, is the blue-eared kingfisher (*pekaka bintik-bintik*) (*Alcedo meninting*) (17 centimetres) which lives near ponds and streams in lowland forest up to 900 metres, feeding on fish, crustaceans and insects and nesting in the banks of forest streams. Its calls are higher, but shorter, than the common kingfisher. A solitary bird, it is shy and secretive, and very difficult to locate.

The white-collared kingfisher (*pekaka sungai*) (*Halcyon chloris*) is a slightly larger bird (20–25 centimetres) with a white breast and collar contrasting with its blue and green wings; the bill and legs are black. It is found mostly in coastal areas and feeds on crabs,

A white-collared kingfisher (*Halcyon chloris*), whose beautiful blue-green colouring reflects the colours of the sea near which it lives.

A common kingfisher (*Alcedo atthis*), a small species of open lowland areas.

A blue-eared kingfisher (*Alcedo meninting*), a tiny species which nests in the banks of forest streams.

A black-capped kingfisher (*Halcyon pileata*), an uncommon migrant sometimes seen in mangrove swamps and paddy fields.

fish, lizards, insects and earthworms. The call is a loud, harsh 'chew-chew-chew-chew-chew'. Its nests are made in a hollow tree near the water's edge or in a termite mound.

A medium-sized bird (28 centimetres), the white-breasted kingfisher (*pekaka belukar*) (*Halcyon smyrnensis*) has a bright blue back and wings, brown underparts and white upper breast and throat; the bill and legs are red. The most common and easily seen of Malaysia's kingfishers, it lives in open country up to 1500 metres. It can be seen perching on electricity wires and on dead branches in paddy fields and oil palm estates with an open view as it searches for prey, including fish, crabs, insects, earthworms, lizards, rats and birds. The call is a loud, chattering 'kilililili'. Its nest is usually a 7-centimetre-wide tunnel about 90 centimetres long with a breeding chamber about 20 centimetres wide in earth banks or termite mounds.

Slightly larger and darker, as well as quieter, than the white-breasted kingfisher is the black-capped kingfisher (*pekaka kopiah hitam*) (*Halcyon pileata*). It has a black head and wings, a white collar, throat and upper breast, rufous abdomen, and deep blue back and tail; its bill and feet are red. An uncommon migrant, this kingfisher is found in mangrove swamps, paddy fields and gardens, as high as 1000 metres. Its diet includes insects, crabs and fish. Nests in river banks are about 60 centimetres long.

Largest of the kingfishers in Malaysia is the stork-billed kingfisher (*pekaka paruh pendek*) (*Halcyon capensis*), which is usually seen in pairs. It is common in lowland forests up to at least 1200 metres, along streams and rivers, and is also seen in mangroves, paddy fields and agricultural lands in coastal areas. Easily recognized by its size (35 centimetres), its dull greyish brown head is in sharp contrast to the bright yellowish brown collar and shiny blue back. The large bill is red with a black tip; the feet are also red. More easily heard than seen, its call is a loud 'ke-ke-keke-ke-keee', often repeated in flight. Prey includes fish, frogs, mice, insects, birds, crabs and birds' eggs. Nests are made in a river bank, a termite nest or a hollow tree trunk.

Woodpeckers

Woodpeckers belong to the family Picidae of the order Piciformes. They are arboreal birds, mostly solitary, of varying sizes and colours; some are very brightly coloured, others dull. Most eat only insects and tree sap, but a few also eat fruit and nuts. These birds are specially adapted to brace themselves against a vertical tree trunk, using their stiff tail feathers and their feet. The hard, sharp bill is designed for chipping wood, either in search of insects or for making a nesting hole. Most have a loud, harsh voice. Woodpeckers are usually seen singly, in pairs or in small flocks. All are cavity nesters, making holes in trees, termite mounds or in the ground. Wood chips provide the only lining in tree holes. A clutch consists of 2–8 glossy, round, white eggs. Both male and female birds share the nest-building, incubation and feeding chores.

The rufous woodpecker (*belatuk biji nangka*) (*Caleus brachyurus*) is a medium-sized bird (21 centimetres) with a short crest, rufous plumage with black bars on the wings, which is found in

secondary forest up to 1500 metres and sometimes in mangroves, singly or in pairs.

In a similar habitat, but only to 1000 metres, is the white-bellied woodpecker (*belatuk gajah*) (*Dryocopus javensis*), the second largest woodpecker in Malaysia. It has black plumage with a white abdomen and red crest; the male has a red forecrown and moustache.

Lowland wooded areas, gardens, coconut groves and cultivated areas are all home to the common goldenback woodpecker (*belatuk pinang muda*) (*Dinopium javanense*). It has a golden yellow back, bright red rump, and black and white striped head and black and white chequered breast; the male has a red crest. Nests are made in dead trees, usually coconut palms.

Piculets

Piculets are very tiny, solitary birds which forage for insects in low bushes, usually along forest edges. Their close relationship to woodpeckers is not obvious except when they are drilling into bamboo to make their nesting hole. The smallest (only 9 centimetres), the rufous piculet (*belatuk kecil*) (*Sasia abnormis*), occurs in forests, especially bamboo stands, up to 1300 metres. It is usually seen in forest undergrowth, hopping among branches searching for insects. Both sexes are olive green above and rufous below; the forehead is yellow in males and rufous in females.

Woodpeckers

As well as their sharp beak with which to find food under the bark of trees and also to make nesting holes, woodpeckers have specially adapted feet and tail feathers which enable them to cling to tree trunks. Unlike most birds which have only one claw facing backward, woodpeckers have two claws facing forward and two facing backward. The claws have very sharp talons which can pierce tree bark to ensure a firm grip. The tail feathers of woodpeckers are stiff and so act as a brace to support the bird.

A buff-necked barred woodpecker (*Meiglyptes tukki*), one of the smaller and less colourful Malaysian species of woodpeckers.

A common goldenback woodpecker (*Dinopium javanense*), a colourful species often found in plantations and coconut groves.

Waterbirds

Waterbirds comprise a large number of families occupying the wetlands of Malaysia. Although most are found in mangroves and mud flats on the west coast of Peninsular Malaysia and in Sabah and Sarawak, others are found near bodies of fresh water. More than half of Malaysia's 156 species of waterbirds (including 12 duck species) are migratory birds, some of which stay for several months; others stop only briefly.

Traditional wooden fishing boats of Kelantan and Terengganu on the east coast of Peninsular Malaysia have a mast guard on the port side of the boat to hold the mast and spars in place when not in use. For unknown reasons, this guard is known as a *bangau* (egret). This example has a colourful egret shape, but many others are very different: some feature a leaf design; others, in the past, incorporated figures from the *wayang kulit* (shadow puppet plays).

Adaptations

All the waterbirds have special features for adaptation to their watery environments. Some, like the herons and egrets, have long legs and necks and long, sharp beaks for wading and probing in the water and mud in search of food. Others, such as the ducks, have webbed feet specially adapted for swimming. The rails and crakes have long legs with long, thin claws adapted for walking in watery habitats.

Herons and egrets

These are all long-legged birds with a long neck and a dagger-like bill. Malaysia has 10 heron species and 6 egret species, all found in wetlands, mostly on the mud flats of the west coast of Peninsular Malaysia, feeding on small fish, aquatic animals and insects.

The black-crowned night heron (*pucung kuak*) (*Nycticorax nycticorax*), a largely nocturnal bird (as its name suggests), is the main resident waterbird species in Malaysia. It is usually found along the west coast of Peninsular Malaysia in wooded swamps and mangroves, where it feeds on freshwater fish, rats and snakes. The juvenile colouring of dark brown streaked with buff is very different from the adult's

The milky stork

The milky stork (*burung upeh*) (*Mycteria cinerea*) is a vulnerable species found only along the west coast of Peninsular Malaysia. There are between 60 and 100 milky storks in Malaysia, most along the mud flats of the Larut Matang area in Perak where they feed on worms, snakes and crabs on the surface of the mud flats or dig food from under the surface using their long, sharp-pointed bills. As they need to move to an area with tall trees for breeding, they fly to Kelompang Island, north of Matang, between October and December every year. Here, they breed in large groups, building nests of dried twigs and branches, arranged in big piles between tree branches. Both male and female birds take part in incubating the eggs and feeding the young. Eggs and chicks are never left unattended because of the presence of predators such as the brahminy kite, monitor lizards and civets. A breeding programme has been in operation for a number of years because the ravages of predators have resulted in a reduction in the colonies of milky storks.

black crown and back, grey wings and white underparts. Breeding colonies are large. Nests are made of small, dry twigs in young mangrove trees, coconut palms and bushes.

A very common resident, whose numbers swell in the northern winter with the arrival of migrants, the little heron (*pucung keladi*) (*Butorides striatus*) inhabits mangroves, mud flats, swamps, paddy fields and river banks, feeding on small fish and other aquatic animals and insects. Nests are made low in trees and bushes. Its cry is a harsh, deep croak.

Another heron with both resident and migrant populations is the purple heron (*pucung serandau*) (*Ardea purpurea*) found in swamps and mangroves, standing still at the edge of the water seeking prey.

All the egrets are migrant birds which winter in Malaysia. The three most common species are the great egret (*bangau besar*) (*Egretta alba*), the little egret

Herons

Long-legged birds with sharp beaks for catching fish, herons are often seen in mangroves and mud flats.

Purple heron (*Ardea purpurea*) in flight

Black-crowned night heron (*Nycticorax nycticorax*) in flight

Immature little heron (*Butorides striatus*)

Black-crowned night heron (*Nycticorax nycticorax*)

Purple heron (*Ardea purpurea*)

(*bangau kecil*) (*Egretta garzetta*) and the cattle egret (*bangau kerbau*) (*Bubulcus ibis*), the only species found in freshwater swamps. All are pure white birds. However, as the breeding season approaches, the plumage and legs change colour. The face of the great egret becomes bluish green, its bill changes from yellow to black and its thighs become greenish. In the little egret, the face becomes bluish green and it develops two long plumes on the nape.

Perhaps the best known of the egrets is the cattle egret, a small bird with a shorter neck and bill than the others. It is found in low, cultivated fields, grasslands and marshes where it feeds on insects disturbed by cattle and buffaloes, and is often seen perching on these animals. Before the breeding season, its head, neck, chest and back become brown and its legs turn red.

The head of a walking stick carved by a Mah Meri craftsman in the shape of a duck's head.

Ducks

Of Malaysia's 12 duck species, only the lesser treeduck (*belibis*) (*Dendrocygna javanica*) and the cotton pygmy goose (*Nettapus coromandelianus*) are residents. The lesser treeduck, the most common species, is found on freshwater swamps and old mining pools. It derives its name from its habit of nesting in tree holes.

Other waterbirds

There are many other waterbirds found in pools of water in open country and in the freshwater swamps of Malaysia. These birds usually have long legs with long, thin claws, adapted for walking on grasses and other floating vegetation in ponds, paddy fields and old mining pools. Their long, sharp bills are designed for probing and catching food in shallow water.

Males and females of the white-breasted waterhen (*ruak ruak*) (*Amaurornis phoenicurus*), one of the most abundant waterbirds in Malaysia, are similar in appearance. This bird is found in most types of habitat close to water, such as marshes, canals, paddy fields and mangroves. An omnivorous bird, feeding on both animals and plants, it is protected under the law, but can be hunted for food during May–August.

Easily recognized by its metallic blue plumage, large red bill and frontal shield and long, reddish legs, the purple swamphen (*pangling*) (*Porphyrio porphyrio*) is fairly common in marshes and man-made lakes, such as old mining pools, with dense floating vegetation. It is a resident bird, totally protected under the Protection of Wildlife Act.

The common sandpiper (*Actitis hypoleucos*), a common migrant, brown above and white below, is a solitary bird which can be seen on both the shore and inland bodies of water.

Migration

During winter in the northern hemisphere, between August and April, many waterbirds migrate south following the East Asian Flyway as far as the Australasian regions. Some birds spend their winters in Malaysia and other Southeast Asian countries. Others stop briefly to gather energy before continuing their journey further south. The northward migration takes place in April–May each year.

The common sandpiper (*Actitis hypoleucos*), a migratory shorebird.

The purple swamphen (*Porphyrio porphyrio*) is a most distinctive bird because of its bright colour.

The white-breasted waterhen (*Amaurornis phoenicurus*) often leaves the vicinity of water to hunt for food.

Lesser treeducks (*Dendrocygna javanica*), also known as whistling teals, the commonest ducks in Malaysia, swimming in formation.

1. An agamid lizard (*Gonocephalus liogaster*), one of a colourful group of reptiles which extend their throat pouch in displays to attract a female or to repel an aggressor.

2. Wallace's flying frog (*Rhacophorus nigropalmatus*) does not fly, but rather glides from high in the tree canopy to the forest floor to escape from predators.

3. The red-tailed racer (*Gonyosoma oxycephalum*) is one of Malaysia's many common, brightly coloured, harmless snakes.

4. The largest crocodile species, the estuarine crocodile (*Crocodylus porosus*), lives at the mouth of rivers, often hiding in aquatic vegetation.

REPTILES AND AMPHIBIANS

A red-tailed pit viper (*Trimeresurus popeiorum*), one of Malaysia's eight poisonous pit vipers.

The reptiles and amphibians of Malaysia include members of the crocodile, turtle, snake, lizard and frog families, and range in size from the very tiniest of frogs (smaller than a person's thumbnail) to crocodiles several metres long. But perhaps the most astonishing feature of this group of animals is the number of species which have the ability to glide from tree to tree or from a tree to the forest floor. These remarkable species include snakes, lizards and frogs.

Most Malaysians are familiar with the estuarine crocodile (*buaya tembaga*) (*Crocodylus porosus*), of which there are increasingly frequent sightings near populated areas. However, the other Malaysian member of the crocodile family, the false gharial (*buaya julong-julong*) (*Tomistoma schlegelii*) is very little known. It has a much narrower snout than the crocodile, and eats only fish.

Malaysia has about 140 species of land snakes, but the majority are harmless. However, as well as the poisonous species there is one nonpoisonous species that is rightly feared: the reticulated python (*ular sawa*) (*Python reticulatus*), the world's second longest snake, which kills by constricting prey as large as a cow or goat before swallowing it whole. Malaysia's 17 poisonous snake species include the largest venomous snake in the world, the king cobra (*ular tedung selar*) (*Ophiophagus hannah*), whose bite can quickly prove fatal. Despite its reputation as a killer, the king cobra has long been revered in many parts of the world.

Lizards in Malaysia include a wide range of species, from the small geckos found on house walls and ceilings eating insects to the largest of the four monitor lizard species, the water monitor lizard (*biawak air*) (*Varanus salvator*), which lives in watery habitats and is prized for its skin, which is exported in large numbers.

There are 18 species of non-marine turtles, tortoises and terrapins in Malaysia. Some are wholly terrestrial, others wholly aquatic, and yet others spend time both in water and on land. Some species are kept at temples as a symbol of longevity, while other species are eaten in the belief they provide nourishing food.

Many endemics are among the more than 165 species of frogs and toads in Malaysia, especially amongst those found in Sabah and Sarawak. Survival features in forest species include poison glands in some toads.

Some Malaysian reptiles and amphibians have economic value. Snakes help to control the rat population in agricultural ecosystems, while amphibians feed on insects. The hides of crocodiles, snakes and lizards are used to produce handbags, shoes and other fashion accessories. Some reptiles and amphibians are considered by certain people to be a gastronomic delicacy. These include terrapins, tortoises, pythons and other snakes, monitor lizards and frogs. The eggs of river terrapins are collected for human consumption.

A spiny hill turtle (*Heosemys spinosa*), one of the common semi-terrestrial pond turtles of Malaysia.

The crocodile and the false gharial

Malaysia has only two of the 25 species of the order Crocodylia, the closest living relatives of dinosaurs and birds, found throughout the tropical regions of the world. These are the estuarine crocodile, the largest of all the crocodile species, and the much less aggressive false gharial, a species with markedly different habits and habitats.

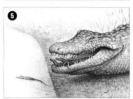

The main distinguishing features of the estuarine crocodile (top) and the false gharial (bottom) are the length of the snout and the sharpness of the teeth. While the crocodile eats a variety of prey, which it tears apart, the gharial's long snout is designed for catching fish, its main food, and its sharp teeth for biting this prey.

A painting of the false gharial (*Tomistoma schlegelii*) from the book by the 19th-century naturalist C. J. Temminck entitled *Verhandelingen over de Natuurlijke Geschiedenis der Nederlandsche Overzeesche Bezittingen.*

The estuarine crocodile

The estuarine crocodile (*buaya tembaga*) (*Crocodylus porosus*) has the most extensive range of all crocodile species; its distribution ranges from eastern India, southern China through Malaysia and the Philippines to northern Australia and the Pacific islands. As its name implies, it inhabits muddy river mouths, particularly the deltas of large rivers, as well as canals near the sea. Fully grown specimens are among the largest living reptiles and are feared as man-eaters. In Malaysia, crocodile attacks in recent years have mostly taken place along the banks of Sarawak rivers.

Groves of nipa palms, which grow in mangroves, are a favourite haunt of the estuarine crocodile. Such habitats are found on the west coast of Peninsular Malaysia and in the huge swampy delta areas of the Sarawak Mangrove Forest Reserve and on the Klias Peninsula, Sabah. The east coast of Peninsular Malaysia is exposed to the strong winds and waves of the South China Sea, so crocodiles take refuge in the middle course of rivers some distance from the sea or in the few sheltered, swampy areas. Crocodiles are also found in the eight major river systems originating from the central mountain range of Peninsular Malaysia, as well as the major rivers of Sabah and Sarawak, which has the largest wild population. Pulau Pinang is the only state without a crocodile population, while Pulau Langkawi is the only offshore island on which they are found.

The largest estuarine crocodile ever caught in Malaysia—5.58 metres long and weighing 1250 kilograms—was captured in early 1997 on the outskirts of Kuala Lumpur when heavy rains caused a former mining pool where it was living to overflow. Considered to be a danger to residents, the crocodile was captured and taken to Melaka Zoo. However, it died a few months later.

Originally a land animal, the crocodile uses its legs only when on land; they are folded against the body when it is swimming. It is the tail which the crocodile uses for swimming, and also to catch prey—which are swept by the crocodile from shallow to deeper water where they are easier to devour. This same method is used in attacking people. The crocodile is itself well protected from predators by the horny scales on its back.

The false gharial

The false gharial (*buaya julong-julong*) (*Tomistoma schlegelii*), the only living species of a very ancient genus, is now found only in southern Thailand, Malaysia and Indonesia. However, excavations in China have revealed that it had a much wider

The estuarine crocodile

The largest and most widespread of the world's crocodiles, the estuarine crocodile (*Crocodylus porosus*) inhabits the mangroves and river estuaries of Peninsular Malaysia as well as of Sabah and Sarawak. At one time considered a threatened species, the number of crocodiles has increased since it was declared a protected species and sightings are no longer rare.

distribution in the past. It is distinguished from the larger estuarine crocodile by its very long snout and sharp teeth, although it does belong to the same family, Crocodylidae. Of similar appearance to the false gharial, but of a different family (Gavialidae), is the 'true' gharial (*Gavialis gangeticus*) of India.

Also distinguishing the false gharial from the estuarine crocodile is its different habitat. Having a preference for fresh water, the false gharial is found in inland swamps and the headwaters of rivers—areas with little siltation or human encroachment. Known locations of gharials include Sungai Tengi (Selangor), Tasik Bera (Pahang) and Sungai Sungkai (Perak). Recent specimens inadvertently captured in fishing nets were from Sungai Pahang and Sungai Bernam (Selangor).

Little is known about the habits of the false gharial. Only in 1994 was the first reported sighting of a false gharial nest made in Sarawak. The nest was a mound about 0.6 metre high and 1.3–1.5 metres in diameter of dry leaves and sticks which had been swept into a pile at the edge of a peatswamp forest under a tree canopy. Submerged in the water about 3–4 metres from the nest, the gharial acted aggressively if people approached the nest. In the nest were 16 eggs, 92–100 centimetres long, with a diameter of 56–59 centimetres and weighing 180–199 grams. All were infertile.

Conservation

The estuarine crocodile and the false gharial were once abundant in Malaysian rivers. However, their numbers have dwindled due to overhunting, land clearing and habitat destruction. Hunting was banned in 1984 and no permits have been issued since then. Due to this and other conservation measures, the population has recovered to the extent that there are now frequent reports of sightings of wild crocodiles, even near Kuala Lumpur.

Legends

The crocodile features in many Malaysian legends as well as in many folk beliefs. It is one of the many animals outwitted by the wily mouse deer. A traditional belief among kampong people is that crocodiles have a guardian spirit. The killing of a person by a crocodile means that the spirit has been offended, and must be appeased by the intervention of a *pawang* or shaman.

Reputedly, a crocodile knocked his victim into the river with a sweep of its tail and then dragged

Crocodile farms

Four crocodile farms have been set up in Malaysia (at Melaka, Pulau Langkawi, Kuching and Sandakan), with the dual aims of catering to the tourist trade and to the demand for exotic reptile skins by the fashion industry. However, as crocodiles are protected species, these farms and their trade in crocodile skins are tightly controlled by the Department of Wildlife and National Parks. A permit must be obtained to set up a crocodile farm, and the reptiles must be obtained from other farms (either locally or overseas). It is forbidden to capture animals from the wild. Similarly, a permit must be obtained for harvesting crocodiles at the farm. Exporters of crocodile skins must have a CITES certificate indicating that the skins are of farmed animals, not of wild specimens. This is an international requirement for trade in products of endangered animals.

Much sought-after (and extremely expensive) fashion goods such as shoes and handbags (above) are made from the skins of the estuarine crocodile, and Malaysian companies are now producing such goods. It is essential that buyers of such goods, particularly foreign visitors (who form a significant percentage of the buyers), also obtain a CITES certificate for their purchases or risk having their prized possessions impounded when taken out of Malaysia.

Visitors to Malaysia's crocodile farms can see crocodiles of all ages, ranging from babies and juveniles to adults.

him down to the bottom until he drowned. The crocodile then returned to the surface, declaring that his victim had drowned, rather than been killed by the crocodile.

Tasik Chini is believed to be home to a sacred white crocodile named Seri Pahang ('the glory of Pahang')—white denotes good luck. At Tasik Bera, the Semelai, an Orang Asli group, have a ban on hunting crocodiles. They believe that their ancestors made a pact with a chief or supreme crocodile living in the lake, promising that the Semelai would not eat crocodile meat if the crocodiles, in turn, promised not to eat Semelai fishermen.

❶

Crocodile motifs

The crocodile is a popular motif in many handicrafts and other cultural items, particularly in Sarawak where the crocodile is held in great esteem by the Iban people.

1. Tin money shaped like crocodiles. Shell-backed ingots, traditionally made at the first smelting of a new tin mine, gradually attained tortoise shape. Other animal shapes were later added and, in addition to their use in magic ceremonies, became a form of currency.

2. Crocodiles are featured in this traditional *pua* weaving from Sarawak.

3. A *moyang* (spirit) carved by the Mah Meri of Peninsular Malaysia depicting the crocodile spirit. Such masks were traditionally used by a *bomoh* in the curing of a patient. In a trance he would decide upon the appropriate moyang for that illness. After the requisite mask had been carved, the bomoh would request the spirit to remove the illness from the patient into itself. The patient was then considered cured, and the mask discarded.

4. An elaborately carved *parang* (jungle knife) handle of deer horn, featuring a crocodile, from Sarawak.

❷

❹ ❸

Tortoises and turtles

Malaysia has 18 species of non-marine tortoises and turtles, which belong to the suborder Cryptodira of the order Chelonia. These species are divided among three families, the truly terrestrial tortoise family, Testudinidae, the semi-aquatic pond turtle family Emydidae, and the aquatic softshell family Trionychidae. Their natural habitat is lowland and hill forests, rivers, estuaries and mangroves. However, due to deforestation, some lowland species have moved closer to populated areas.

The Burmese brown tortoise (*Manouria emys*), Malaysia's largest tortoise, lays its spherical eggs among leaf litter on the forest floor. It floats if placed in water, where it cannot survive.

The Malayan snail-eating turtle (*Malaemys subtrijuga*) is found only in northern Peninsular Malaysia, near the border with Thailand. Little is known about this turtle, except its fondness for eating snails.

Tortoises

The three Malaysian tortoises of the family Testudinidae are exclusively terrestrial in habit, and are all forest dwellers. The most obvious characteristics are an arched carapace (shell) and a stubby head with prominent scales. Their legs are club-footed and have thick, heavy scales, while the claws are large and strong. Their feet are not webbed, so they cannot swim well. The Burmese brown tortoise (*baning perang*) (*Manouria emys*), which is the largest of the Asian tortoises with a shell length of at least 50 centimetres, is common. The yellow-headed tortoise (*baning lonjong*) (*Indotestudo elongata*) is less common, and the impressed tortoise (*baning bukit*) (*Manouria impressa*) is very rare. Both are of medium size, measuring less than 25 centimetres.

Pond turtles

The pond turtles of the family Emydidae are the most diverse family, with 11 species known to occur in Malaysia. They are all semi-terrestrial in habit. The carapace is not arched, but rather concave. They are distinguishable from the tortoises by their relatively short, but not stubby, limbs. The claws are long and sharp; the toes are longer and partially webbed, and the turtles are good swimmers. Most are of medium size, with a shell length of 18–45 centimetres, seldom attaining more than 50 centimetres.

Largest are the giant Asian pond turtle (*juju-juku besar*) (*Olitia borneensis*) and the giant hill turtle (*kura besar*) (*Heosemys grandis*) which can exceed 80 centimetres in length. The most common are the Malayan box turtle (*kura katup*) (*Cuora amboinensis*), the spiny hill turtle (*kura duri*) (*Heosemys spinosa*), the black marsh turtle (*jelebu hitam*) (*Siebenrockiella crassicollis*) and the temple turtle (*Hieremys annandalei*). These common species are found on forest floors, in forest streams, plantations and paddy fields.

Three species, the Malayan snail-eating turtle (*jelebu siput*) (*Malaemys subtrijuga*), the Malayan flat-shelled turtle (*biuku*) (*Notochelys platynota*) and the Asian leaf turtle (*kura daun*) (*Cyclemys dentata*) are considered rare, and have been placed on the endangered list.

A selection of Malaysia's non-marine turtles and tortoises

1. Burmese brown tortoise (*Manouria emys*)
2. Giant hill turtle (*Heosemys grandis*)
3. Spiny hill turtle (*Heosemys spinosa*)
4. Yellow-headed tortoise (*Indotestudo elongata*)
5. Malayan box turtle (*Cuora amboinensis*)
6. Black marsh turtle (*Siebenrockiella crassicollis*)
7. Asiatic (mud) softshell (*Amyda cartilagineus*)
8. River terrapin (*Batagur baska*)

Included in the family Emydidae are two species of truly aquatic forms, the river terrapin (*tuntung sungai*) (*Batagur baska*) and the painted terrapin (*tuntung laut*) (*Callagur borneoensis*), which live in rivers, estuaries and mangrove areas. Although similar

DISTINGUISHING FEATURES OF TORTOISES, POND TURTLES AND SOFTSHELL TURTLES			
	TORTOISES	**POND TURTLES**	**SOFTSHELL TURTLES**
Habitat	terrestrial	semi-terrestrial/aquatic	aquatic
Location	forest	forest, plantations, paddy fields, rivers, estuaries	major rivers and estuaries
Head	stubby with prominent scales	long neck	two species with short snout; two species with long snout
Carapace	arched	concave	flattened, disc-like
Legs	large, club-footed, with thick scales and large claws; not webbed	short but not stubby; long, sharp claws; partly webbed	semicircular, paddle-like
Length (cm)	25–50	18–50	20–100
Diet	mainly herbivorous	mainly herbivorous	carnivorous

in appearance, the painted terrapin is smaller, with a maximum shell length of about 50 centimetres. The river terrapin may reach 60 centimetres. Also, the painted terrapin has five toes on the forefoot, while the river terrapin has only four.

The breeding activity of these two species of turtles has been studied in detail. The former lays eggs about three times per season, with an average of 24 eggs each time, and the incubation period is 66–68 days. The latter lays eggs twice per season, with an average of 12 eggs and an incubation period of 55–75 days. Because of the rapid depletion of their habitat through pollution and development, breeding programmes have been established for both species.

Softshell turtles

The softshell or freshwater turtles are easily distinguished by their flattened disc-like shell and semicircular, paddle-like limbs. Malaysia has four species of these turtles, which are carnivorous and feed on fish and aquatic snails. The Malayan softshell (*labi Melayu*) (*Dogania subplana*), the smallest with a shell length of less than 20 centimetres, is distinguished by four distinct black spots on the carapace. It is commonly found in forest streams.

The other three species, the Asiatic giant softshell (*labi besar*) (*Pelochelys bibroni*), the Asiatic/mud softshell (*labi biasa*) (*Amyda cartilagineus*) and the narrow-headed softshell (*labi bunga*) (*Chitra chitra*), grow to prodigious proportions with a shell length of 70–75 centimetres, with occasional sightings exceeding a metre in length in the mud softshell species. With the exception of the narrow-headed softshell, the rarest of all, these turtles are confined to major rivers and estuaries.

The painted terrapin

The most colourful of the Malaysian turtles, the painted terrapin (*tuntung laut*) (*Callagur borneoensis*) is unique in that the female swims from the river to the sea, laying her eggs on the beach in a manner similar to sea turtles. However, eggs of the painted terrapin can easily be distinguished from those of sea turtles as they are larger (66–68 centimetres long), and oval in shape rather than round.

1. The dark phase (non- breeding colour) of the male painted terrapin.
2. The intermediate phase (between non-breeding and breeding colours) of the male painted terrapin.
3. The light phase (breeding colour) of the male painted terrapin.
4. A female painted terrapin laying eggs on the beach at Kuala Setiu, Terengganu. *INSET*: Eggs of the painted terrapin.

While the female has a brown head and shell, the male painted terrapin has two colour phases, dark during the non-breeding season and light during the breeding season. In the dark phase, the male has a dark grey or black head with a dull orange stripe running between the eyes to the tip of the snout. The shell is drab brown. During the light phase, the male has a pure white head, a scarlet stripe between the eyes and the tip of the snout is bluish. The shell becomes nearly white, with three black stripes.

Such dramatic colouring has given the painted terrapin a role in Malaysian folklore. Egg collectors in Kedah have told the story of how the annual nesting migration of the painted terrapin females is led by the Raja Tuntung, a magical species with a bright white head and a scarlet stripe between the eyes.

People and turtles

Some common species of tortoises and turtles are kept as pets, while others are used as offerings at religious ceremonies and then released back to the wild. The practice is believed to enhance one's longevity. Ponds of tortoises are a common sight at Buddhist temples in Malaysia.

When cooked with herbs, turtle meat is said to have aphrodisiac value. Because of the demand for turtle meat, a number of species are threatened, especially the softshell turtles such as the mud softshell and the Malayan softshell, but a few of the aquatic species of pond turtles such as the box turtle and the black marsh turtle are also in danger.

Two decades ago, the softshell species were a common sight in lakes, mangroves, swampy areas, ponds, rivers and forest streams. It is, however, rare to sight them today. Little is known of their biology, and no attempts have yet been made to breed them in captivity. Breeding programmes have only been established for the river terrapin and the painted terrapin.

Tortoises at the Sam Poh Tong Buddhist Temple, a cave temple near Ipoh, Perak. To the Chinese, the tortoise is a symbol of longevity and is often released after religious ceremonies.

Lizards

Worldwide, some 3,500 lizard species in more than 300 genera are recognized today. Malaysia has 143 lizard species: 94 in Peninsular Malaysia and 82 in Sabah and Sarawak, divided into 9 families, of which skinks (Scincidae), geckos (Gekkonidae) and agamid lizards (Agamidae) are the most diverse and well represented.

The tokay gecko (*Gekko gecko*), one of the biggest geckos in Malaysia, is one of two forest species which also enter homes.

There are legendary tales about the gecko's tail, such as that if it breaks off and wriggles into your ear, you will be deaf. One fact about the gecko's tail remains. The tail serves as an excellent organ for defence when the gecko is threatened. It autotomizes easily at a specific fracture plane (above) and the wriggling motion of tho chod tail distracts the predator while the gecko makes its escape, as the *Cyrtodactylus malayanus* (below). The tail is eventually regenerated, albeit smaller than the original. The regenerated tail is sometimes forked.

Geckos

Geckos (*cicak*) are lizards with cylindrical, flattened bodies and large heads and eyes. The pupil of a gecko's eye not only indicates the lizard's feeding style, but also its habits; nocturnal forms have vertical slit pupils, while diurnal species have round pupils. Nocturnal geckos are 'sit-and-wait' predators, whereas diurnal geckos hunt for their prey. Most geckos have no eyelids; instead, the eyes are covered with a transparent spectacle, replaced each time the gecko moults.

The lizards which stalk insect pests on walls and ceilings of houses are the house geckos; most familiar are the common house gecko (*Hemidactylus frenatus*), flat-tailed gecko (*H. platyurus*), four-clawed gecko (*Gehyra mutilata*) and spotted house gecko (*Gekko monarchus*).

Forest-dwelling species live on trees and rocks, preying on crawling insects. Two forest geckos which sometimes invade homes are the tokay gecko (*Gekko gecko*) and the large forest gecko (*G. smithi*). They are the largest living geckos in Malaysia and can attain a total length of about 35 centimetres.

Geckos are unique as the only group of lizards with vocal sacs, giving them a true voice. Vocalizations, associated with their nocturnal habits, are mostly monosyllabic, and both lungs and larynx are involved in sound production. The tokay gecko is often heard producing a crackling and 'tuk-ko-tuk-ko' sound in the evening, and occasionally also during the day.

Most geckos lay two round, white, hard-shelled eggs. The newly laid eggs have a sticky liquid which enables them to adhere to one another and to any available surface.

Geckos use their long tongue to clean their eyes, which are covered by a transparent spectacle; they have no eyelids.

Skinks

Skinks (*bengkarung*) form the largest family amongst lizards, but they are least known because of their secretive habits. The typical skink is a small, short-legged lizard with a neck of the same thickness as the narrow head, and a long, cylindrical body, usually brownish. Its smooth scales have a clear metallic lustre. Skinks are unique among lizards in having a secondary palate. Unlike geckos, skinks have movable eyelids. However, like geckos, the tail detaches easily. Most skinks lay eggs, but some bear live young.

Skinks have wide habitat preferences. The most common, the common sun skink (*Mabuya multifaciata*), is an active ground lizard often heard scuffling in the leaf litter of the forest floor; it is frequently seen basking in sun spots and may sometimes be found even in rural gardens.

Agamid lizards

The tropical rainforests of Malaysia are home to some of the most colourful, cryptic and ornamented lizards. These are small to medium-sized, strongly built, tree-dwelling lizards of the family Agamidae. Nearly all agamids have the ability to change body colour rapidly from shades of green, yellow or blue to a dark colour for signal communication during courtship displays, aggression or camouflage. For this reason, they are often, incorrectly, called chameleons.

The habits of agamid lizards are quite diverse. One genus, the common flying lizard (*Draco* spp.), is arboreal; the others, commonly called tree lizards, are semi-arboreal. Agamids are all insectivorous; they use the claws on their toes to perch on tree trunks and branches to prey on moving insects. The forest

The gecko's foot

The remarkable ability of geckos to walk on vertical surfaces and even upside down on ceilings comes from the underside of the feet, which have overlapping flaps of skin called lamellae, as in *Gekko smithi* (right). Each lamella is covered with closely set microscopic projections called setae; a single toe pad has as many as a million setae. The clinging mechanism comes from the adhesive forces between molecules at the terminal surfaces of the setae and the surface of contact. As the gecko moves, it adjusts its digital pads to conform to any surface irregularities to maximize the number of setae in contact with the surface. Thus, the gecko uses the laws of physics and its unique digital specialization to walk in any orientation on almost any surface.

Agamid lizards

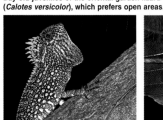

Day-old juveniles of the oriental garden lizard (*Calotes versicolor*), which prefers open areas.

An oriental garden lizard (*versicolor*) in breeding col

An anglehead lizard (*Gonocephalus chamaeleontinus*), found in Malaysia only on Pulau Tioman.

A bush agamid (*Phoxophrys nigrilabris*), found in Sabah but not in Peninsular Malaysia.

A common skink scuffling in the leaf litter of the forest floor on Pulau Tioman, off the east coast of Peninsular Malaysia.

Monitor exploitation

Monitors are protected in Malaysia by the Protection of Wildlife Act (1972) and are listed under Appendix 2 of CITES. The water monitor is the most heavily exploited monitor. It is killed for the skin, which is a source of ornamental leather, as well as the meat, but is not threatened with extinction. The annual international trade in water monitor skins in Southeast Asia is estimated at 1–1.5 million skins. Besides this commercial exploitation, monitor lizard meat and eggs are a source of food for indigenous peoples, who use the fat and other animal parts for medicinal purposes. Despite the commercial trading of monitor lizards (for their skin and meat) and agamids (as exotic pets), these species and their relatives are presently not threatened by human exploitation.

The clouded monitor lizard (*Varanus bengalensis nebulosus*), the only one of Malaysia's four monitor lizards which is restricted to Peninsular Malaysia.

agamids are territorial. A male agamid accepts more than one female into his territory, but drives away other males. Individuals also move from one tree to another and, on rare occasions, even forage on the forest floor. Whether arboreal or semi-arboreal, these lizards deposit their eggs in a nest in the soil which they dig with their forelimbs. The clutch size varies from 2 to 20 eggs, depending on the species and body size; the oriental garden lizard (*Calotes versicolor*) has the largest clutch. Agamid eggs are enveloped by a soft, parchment-like eggshell which is permeable to water and atmospheric gases. Once a clutch is laid and the nest covered with soil, the female leaves.

The male grand anglehead lizard (*Gonocephalus grandis*) is ornamented with tall, sail-like crests of large, overlapping, bluish green scales from the crown of the head to the neck, and from the shoulders to the tail, making it look like a tiny dinosaur. Females have no crest and have different body markings. Only found on Pulau Tioman, *Gonocephalus chamaeleontinus* is a magnificent lizard with a lichen green body, a prominent crest and a large gular pouch.

The monitor lizard uses its long, retractable, forked tongue to detect smells in the environment.

The most common and conspicuous of all agamid lizards near human settlements is the oriental garden lizard (*Calotes versicolor*), which prefers open areas, not closed canopy forests. The males are crimson red on the head and upper half of the body

during the breeding season. A very adaptable and aggressive lizard, it seems to outcompete the green-crested lizard (*Bronchocela cristatella*) when their ranges overlap.

Monitor lizards

Monitor lizards (*biawak*) belong to the family Varanidae, with one genus, *Varanus*. Largest and best known is the komodo dragon (*V. komodoensis*) of Indonesia. In Malaysia, the largest of these lizards is the water monitor (*biawak air*) (*V. salvator*), which grows up to 3 metres long and 25 kilograms in weight. The other three Malaysian monitors are the Dumeril's monitor (*biawak kudung*) (*V. dumerilii*), the rough-neck monitor (*biawak serunai*) (*V. rudicollis*) and the clouded monitor (*biawak bunga*) (*V. bengalensis nebulosus*), which is not found in Sabah and Sarawak.

Adult monitors are light to dark brown to almost black in colour, but juveniles often have a distinctive pattern of light yellow spots. Their elongated bodies are supported by sturdy limbs ending in strong, curved claws. The loose skin around the long neck can be inflated when the monitor is threatened. Although they have sharp, evenly spaced teeth with which to seize their prey, monitors cannot chew with them. Instead, victims are swallowed whole. Often used as a weapon to lash out at enemies, the tail is long and muscular, moving about in a snake-like manner to propel the body forward when the monitor swims. The nostrils are located near the end of the snout, facilitating breathing while swimming.

The most numerous is the water monitor, which occurs in riverine and other habitats near water not disturbed by human activities. The clouded monitor ranges from the edge of high jungle into more open plantations, orchards and dry wasteland. When chased by predators or hunters with dogs, it runs fast for a short distance before disappearing up a tree.

Monitors are carnivorous and hunt for live prey such as crabs, insects, birds, fish, amphibians and rodents. They may raid poultry farms to feed on live birds and eggs and also excavate turtle and crocodile nests for eggs. Adult monitors dig burrows to use as sleeping dens as well as refuge from predators. Female water monitors lay about 20 eggs per clutch in burrows dug out by the river bank.

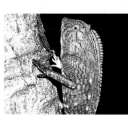

An adult male oriental garden lizard (*Calotes versicolor*).

Gonocephalus abbotti, from the Temengor Forest Reserve.

The female grand anglehead lizard (*Gonocephalus grandis*) lacks the dramatic colouring of the male.

A highland anglehead lizard (*Gonocephalus robinsonii*) from Cameron Highlands, Pahang.

Lizard motifs

❶

❷

1. In Sarawak, monitor lizards with a forked tail are a popular handicraft motif, as in this one created from shells.

2. Although seldom added to the roofs of newly constructed Malay houses in villages, carvings of lizard heads were traditionally a common decoration. A selection of such carvings is shown here.

Flying reptiles

Flight adaptation in reptiles, which has existed since the time of the Pterosaurs during the Triassic period about 225 million years ago, is now confined to a very few lizards and snakes. In Malaysia, these are geckos of the genus Ptychozoon, *lizards of* Draco *species and snakes of the genus* Chrysopelea, *whose ability to glide from tree to tree is an adaptation to life in the dense rainforest.*

MALAYSIA'S MOST COMMON FLYING REPTILES	
COMMON NAME	**SCIENTIFIC NAME**
Flying geckos	
Horsfield's flying gecko	*Ptychozoon horsfieldi*
Kuhl's flying gecko	*Ptychozoon kuhli*
Flying lizards	
Spotted-winged flying lizard	*Draco melanopogon*
Common flying lizard	*Draco volans*
Giant-sized flying lizard	*Draco maximus*
Multicoloured flying lizard	*Draco quinquefasciatus*
Flying snakes	
Paradise tree snake	*Chrysopelea paradisi*
Twin-barred tree snake	*Chrysopelea pelias*
Golden tree snake	*Chrysopelea ornata*

Flight

Unlike birds, flying reptiles are not capable of true flight. They can only glide, making use of the membranes or other structural adaptations of their body. Together with their camouflage coloration, these adaptations serve as a means of escape from predators, and also enhance hunting mobility. The survival success of these flying reptiles is higher than that of the strictly terrestrial species which must be constantly on guard against ground-dwelling predators. Birds of prey are the only likely predator of the flying reptiles.

Flying geckos

Malaysia's two flying geckos (*cicak terbang*) of the genus *Ptychozoon* (family Gekkonidae), Horsfield's flying gecko (*P. horsfieldi*) and Kuhl's flying gecko (*P. kuhli*), are small geckos (12–15 centimetres long) distinguished by their broadly webbed feet and the skin flaps on the sides of the head, body and tail which act as a parachute when the gecko is gliding.

Their colour is usually grey or brown with darker markings, but they can change colour according to the habitat. This dull colouring, as well as their skin flaps, makes a gecko resting on a tree trunk indistinguishable from a patch of fungus, and thus

Perfectly camouflaged by its dull coloration, a flying gecko (*Ptychozoon kuhli*) flattened against a tree trunk resembles a patch of fungus.

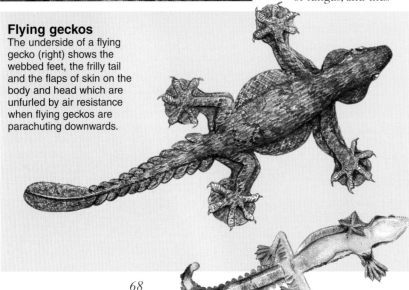

Flying geckos

The underside of a flying gecko (right) shows the webbed feet, the frilly tail and the flaps of skin on the body and head which are unfurled by air resistance when flying geckos are parachuting downwards.

Flying lizards

In order to glide from tree to tree, flying lizards spread their large wing-like flying membrane. The skeleton of a flying lizard (above right) clearly shows the ribs which the lizard extends to support this membrane.

When resting on a tree trunk. flying lizards fold their flying membrane against the body and are very difficult to see.

easily overlooked. Although arboreal by nature (even laying their eggs on the bark of a tree), flying geckos sometimes descend to the ground to feed on insects.

The skin flaps on the flying geckos cannot be spread by muscular action, but are raised by air resistance. Together with the webbed feet and the frilled tail, they act as a planing surface for the gecko to parachute downwards at a steep angle to the forest floor in a similar manner to the flying frogs.

The diet of flying geckos consists mainly of ants, beetles and moths. Little is known about their breeding habits. In captivity, eggs laid in clutches of two stuck to the bark of trees took 45–60 days to hatch. It is assumed that hatchlings cannot glide like adults as their skin flaps are not well developed.

Two eggs of a flying gecko (*Ptychozoon kuhli*) laid on the bark of a tree.

Flying lizards

Malaysia has 10 species of flying lizards (*cicak kubin*), small reptiles (20–40 centimetres long) belonging to the genus *Draco* of the family Agamidae, which are widely distributed and are mainly forest species. Commonest is the spotted-winged flying lizard (*D. melanopogon*), though perhaps more frequently seen is the common flying lizard (*D. volans*), as it can also be found outside the forest in home gardens. Largest is the giant-sized flying lizard (*D. maximus*), while the most colourful is the multicoloured flying lizard (*D. quinquefasciatus*). These exclusively arboreal and diurnal reptiles have a flat body with a long tail and long legs. Unlike the flying geckos, the toes of the flying lizards are not webbed.

especially the paradise tree snake (*C. paradisi*) which is black with a green spot on each scale and a row of bright red four-petalled spots along the middle. The twin-barred tree snake (*C. pelias*) is brick red with pairs of black crossbars, each enclosing a greenish yellow bar, with the brightest colours on the forepart of its body, while the golden tree snake (*C. ornata*) is uniformly green, with each scale bordered with black and bisected longitudinally by a black line.

Flying snakes glide through the air without the skin flaps of the flying geckos or the 'wings' of the flying lizards, but have a unique system of their own. The body is flattened dorso-ventrally and hollowed out below so that it is hemicylindrical, similar to half of a split bamboo, giving it the ability to glide in a controlled manner, in a similar method to the flying lizard. In preparing to glide, a snake hangs head downward with its body in a wide S-shape. Upon launching itself from a branch, it hollows its belly between the ridges on the ventral side while keeping the concave surface of its body downwards, as a parachute does, thus buoying the body and retarding its speed of descent. The gliding distance is usually 20–30 metres, but may be as far as 50 metres.

Mostly forest species, flying snakes are widely distributed in Malaysia. The paradise tree snake, however, may also be found in house gardens, and even inside houses, in areas close to forest, to feed on house geckos or lizards.

Arboreal and diurnal in habit, the nonpoisonous flying snakes are graceful and agile climbers, moving from branch to branch or up a tree trunk with ease, with the added ability to spring horizontally and upwards so as to glide from a height to the ground, or to another tree, feeding mostly on lizards. Little is known of their breeding habits. Eggs of a captive paradise tree snake took 35–41 days to hatch and were laid in a nest on the ground. Unlike the adults, the young are not able to glide from tree to tree.

Flying snakes

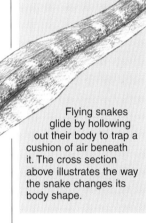

Flying snakes glide by hollowing out their body to trap a cushion of air beneath it. The cross section above illustrates the way the snake changes its body shape.

The large, lateral, wing-like membrane which enables gliding is supported by five or six long ribs; stretched out, this membrane resembles a fan. There is a small crest on the head. Like the other agamid lizards, the *Draco* species have a throat projection (gular appendage); they also have a lateral flap of skin on each side. The gular pouch of the male is larger than the female's as it is inflated for use in courtship rituals. While the coloration of the lizards resembles the bark of the trees on which they live, the bright colours of the gular pouch are an example of flash coloration, designed to draw attention to the animal.

With the aid of the fully extended wing membrane, the flying lizards glide from tree to tree, usually from a higher to a lower elevation; the angle of descent is about 20–30 degrees. The gular pouch is fully extended and is used as a rudder. When landing on a tree, the head is pointed upwards and the lizard runs up the trunk in a jerky manner. The length of the glide depends on the height from which it is launched, and is assisted by the wind currents; it can range from 3 to 6 metres. When the lizard is at rest on a tree trunk, the flying membrane is closely folded against the body, and it is almost invisible.

According to an Orang Asli legend, if one is bitten by a flying lizard at midday, death is imminent. Thus, few Orang Asli are willing to handle live flying lizards.

Flying snakes

In Malaysia, flying snakes are represented by three species of a single genus, *Chrysopelea*. Slender snakes about 1 metre long, they are very colourful,

The paradise tree snake (*Chrysopelea paradisi*), the most colourful of Malaysia's flying snakes.

Poisonous snakes

The legendary king cobra (Ophiophagus hannah), reputedly the world's largest venomous snake, is one of only 17 poisonous land snake species in Malaysia. These snakes belong to two families, the Viperidae containing eight pit vipers, the only family of snakes in the world with heat-sensing pits for detecting prey in the dark, and the Elapidae, with two cobras, three kraits and four coral snakes.

Fangs of the viper and cobra

The fangs of pit vipers (above) are long, movable and very prominent. When the mouth is closed, the fangs lie flat, with the points directed backwards. In contrast, the fangs of cobras (below) (and of kraits and coral snakes) are smaller and fixed.

Pit vipers

Among the eight species of Malaysian pit vipers, those frequently encountered are the Malayan (marbled) pit viper (*Calloselasma rhodostoma*), Wagler's pit viper (*Tropidolaemus (Trimeresurus) wagleri*) and the shore (mangrove) pit viper (*Trimeresurus purpureomaculatus*). The other species are by no means rare, but are encountered less often as they are confined to lowland and highland forests.

The Malayan pit viper is the most dangerous, causing the majority of snake bites in northern Peninsular Malaysia, where it is most abundant. It is closely associated with areas near human habitation, such as paddy fields and rubber plantations; hence the large number of bites. In other parts of Peninsular Malaysia and also in Sabah and Sarawak, viper bites are uncommon. They are usually caused by the shore pit viper, as it is very common in mangrove forests which are frequented by woodcutters.

Most pit vipers are arboreal in habit; only the Malayan pit viper and mountain pit viper (*Ovophis (Trimeresurus) monticola*) are terrestrial. Equally active during the day and night, they feed on warm-blooded animals such as rats, birds and occasionally frogs and lizards. Hunting is aided by their unique heat-sensing pits. The venom of the Malayan pit viper has anti-clotting properties which act as an anticoagulant for blood clotting in the heart vessels.

Snakebite marks

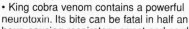

king cobra

krait

cobra   sea snake   coral snake

viper   nonpoisonous snake

Snakebites

The snakebite patterns of Malaysian snakes can help in identifying a snakebite. Such identification is important as the correct treatment for snakebite varies according to the species, and if the snake cannot be caught identification has to be made from the bite marks. Nonpoisonous snakes have four rows of teeth marks, while poisonous snakes have only two, with two distinct fang marks.

When a snakebite occurs, the following measures should be taken, as quickly as possible:
1. Reassure the patient, and keep him warm.
2. Inspect the wound and try to establish if the bite is that of a poisonous snake.
3. Immobilize the patient, particularly the bite area (if a limb, use a splint and keep the limb below the level of the heart).
4. Clean the site of the bite, but do not cut or try to suck out the venom.
5. Apply a tourniquet.
6. Take the patient to the nearest hospital.
7. Kill the snake and take it for identification so the correct antivenene can be given.

Distinctive characteristics of the four families of poisonous snakes in Malaysia

Cobras

- The only snakes which have a hood, which is formed by loose skin of the neck being pushed out by the ribs, which are elongated in that part of the body. The hood is raised when the snake is alarmed. Even hatchlings raise their hood if disturbed.
- The black cobra hisses while squirting venom, which it can do accurately up to a distance of 1.5 metres.
- The king cobra is dangerous because of its size and the amount of venom.
- The king cobra builds a nest of grass and leaves, lays her eggs, and covers them. She does not sit on the nest, but stays nearby, guarding against predators for two months until the eggs hatch. The eggs of the black cobra are brooded by both sexes; no nest is made.
- King cobra venom contains a powerful neurotoxin. Its bite can be fatal in half an hour, causing respiratory arrest and cardiac failure.
- Black cobra venom can kill in 1–6 hours if the victim is not treated.

The common black cobra (*Naja sumatranus*)

Pit vipers

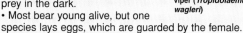

- The head is triangular, and is distinct from the neck because of the poison glands and associated muscles.
- The only snakes with a pair of loreal pits between the eyes and nostrils which are capable of detecting radiant heat in minute amounts. They are used by the snake for locating warm-blooded prey in the dark.
- Most bear young alive, but one species lays eggs, which are guarded by the female.
- The venom is haemorrhagic. A viper bite is painful and causes localized swelling.

A juvenile Wagler's pit viper (*Tropidolaemus wagleri*)

Kraits
- All are brightly coloured snakes; two are banded, and the third has a red head and tail.
- The head is slightly distinct from the body, and the pupil of the eye is round.
- All have a mild, sluggish temperament.
- All lay eggs.
- All are nocturnal and terrestrial snakes.
- The venom is extremely potent, and acts as a neurotoxin.
- All are triangular in cross section.
- All eat other snakes, lizards and frogs.

The banded krait (*Bungarus fasciatus*)

Coral snakes
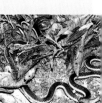

- All are small, brightly coloured, burrowing snakes which are mainly nocturnal.
- All are non-aggressive unless roughly handled.
- All lay eggs.
- Their small mouth is generally believed not capable of biting an adult.
- The venom glands extend back for about one-third or more of the snake's body length.
- The defence mechanism of the blue Malayan coral snake is to tuck its head under its coils and raise its tail to show its coral red belly.

The blue Malayan coral snake (*Maticora bivirgata*)

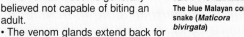

The legendary cobras

Myths have surrounded cobras, among the world's most poisonous snakes, since ancient Egyptian times.

In India, cobras are worshipped by some ethnic groups. During the Nagpanchami festival, cobras are taken from house to house and offered milk and money.

In Buddhist mythology, a cobra protected the Lord Buddha from the sun by raising its hood over his head. The Lord Buddha blessed the snake by placing his index and middle fingers on the snake's head, leaving an indelible pair of 'eyes'.

It is also believed that when a cobra is killed, its mate will appear to take revenge, but there is little evidence to show that this belief is true.

The king cobra (*Ophiophagus hannah*)

The cobra is perhaps best known for its connection with snake charmers, who have the reputation of being able to 'charm' snakes into swaying with the melody from the snake charmer's flute. The reality is that the snakes 'dance' to the rhythmical movement of the snake charmer and his flute as, like all snakes, cobras are deaf to airborne noises.

A traditional Malay belief is that the king cobra has a bright stone (*gemala*) in its head, the radiance of which makes it visible even on the darkest night.

In Malaysia, the meat of the cobra cooked with medicinal herbs is favoured by some in the Chinese community for its supposed aphrodisiac value. It is also believed that eating the raw gall bladder mixed with wine cleanses one's blood and aids in achieving a smoother complexion. However, there is no scientific basis for such beliefs.

Cobras, kraits and coral snakes

These are snakes of the family Elapidae. They are distinguishable from other land snakes by their enlarged poison fangs at the front of the upper jaws. Another characteristic of these snakes is the enlarged third upper labial (upper lip) which touches the eye in the cobra species. In the kraits and coral snakes, the third and fourth upper labials touch the eye. The snakes of this family are represented by five genera with nine species in Malaysia. These are two species of cobras, three of kraits and four of coral snakes.

Among the elapid snakes, the king cobra (*Ophiophagus hannah*), the common black cobra (*Naja sumatranus*), the banded krait (*Bungarus fasciatus*) and the blue Malayan coral snake (*Maticora bivirgata*) are encountered more often than the other species. The king cobra is the largest venomous snake in the world. It can reach 6 metres in length, and is uniformly yellowish brown in colour, while the common black cobra is glossy black and is about 1.5 metres long. The banded krait is banded black and white and is about 1 metre long, while the blue Malayan coral snake is dark blue with a red head, tail

and belly. It is a long and slender snake, which grows up to 1.5 metres in length.

The common black cobra and banded krait are the two most dangerous species due to their close association with human surroundings, such as gardens, fields and even homes. Most elapid snake bite cases are caused by these two species; a higher number of bites are caused by the common black cobra, which also causes the most fatalities among snake bite victims. Although the king cobra has more powerful venom, it lives further from habited areas, and so does not cause as many fatalities.

Cobras are predators of rats, and also eat other vertebrate animals such as frogs and lizards. In oil palm plantations infested with the wood rat (*Rattus tiomanicus*), a major pest, cobras are commonly found due to the availability of food resources.

An adult Wagler's pit viper (*Tropidolaemus wagleri*), whose bright colouring is very different from that of the juvenile of the species.

Many temples in Malaysia keep snakes as symbols of good luck and prosperity. At the Penang Snake Temple, the deity Chor Su Kong is surrounded by poisonous Wagler's pit viper snakes. The deity's birthday on the sixth day of the first moon of the Chinese lunar calendar is celebrated on a large scale; worshippers bring offerings such as eggs for the vipers.

MALAYSIA'S POISONOUS LAND SNAKES

COMMON NAME	MALAY NAME	SCIENTIFIC NAME
Cobras		
Common black	ular senduk	Naja sumatranus
King	ular tedung selar	Ophiophagus hannah
Kraits		
Banded	ular katang belang	Bungarus fasciatus
Malayan	ular katang tebu	Bungarus candidus
Red-headed	ular katang kepala merah	Bungarus flaviceps
Pit vipers		
Flat-nosed	ular kapak hidung pipeh	Trimeresurus puniceus
Hagen's	ular kapak Hagen	Trimeresurus hageni
Malayan (marbled)	ular kapak bodoh	Calloselasma rhodostoma
Mountain	ular kapak gunung	Ovophis monticola
Red-tailed	ular kapak ekor merah	Trimeresurus popeiorum
Shore (mangrove)	ular kapak bakau	Trimeresurus purpureomaculatus
Sumatran	ular kapak Sumatera	Trimeresurus sumatranus
Wagler's	ular kapak tokong	Tropidolaemus wagleri
Coral snakes		
Banded	ular pantai belang	Maticora intestinalis
Blue Malayan	ular pantai biru	Maticora bivirgata
Small-spotted	ular pantai bintik kecil	Callophis maculiceps
Spotted	ular pantai bintik	Callophis gracilis

Nonpoisonous snakes

Among the 144 species of nonpoisonous snakes (ular) of Malaysia, belonging to six families, are many which play an important role in controlling the rat and snake populations in both forests and cultivated areas. They range from the smallest worm-like blind snakes about 15 centimetres long to the second largest snake in the world, the reticulated python, which can reach 10 metres. Some are very colourful; the coloration of other species blends with their habitat.

In the most common type of snake movement—serpentine movement (above)—the snake presses its body against the ground, producing both sideways and backwards thrust. Speed is increased by increasing the size of body waves. The use of muscles to erect the scales increases the friction. A modified form of the serpentine movement is the concertina movement (below), which is often used in burrows and tunnels.

Blind snakes

Members of the family Typhlopidae comprise eight species of small worm-like snakes; their eyes are hidden under the scales of the head and thus hardly visible. They are burrowing snakes, usually found in loose soil, under rocks or fallen trees. Strictly nocturnal, they emerge at dusk to forage for food—insects and their larvae. Only the smallest, the black common blind snake (*Ramphotyphlops braminus*), is found near houses; it is often mistaken for an earthworm.

Pipe snakes

Similar in appearance to the blind snakes, the two pipe snakes (family Uropeltidae) are also burrowing species. But they are larger than the blind snakes, and their eyes are visible. Leonard's pipe snake (*Anomochilus leonardi*) is a rare forest species, while the red-tailed pipe snake (*Cylindrophis rufus*), a common species, is also found outside the forest. Both have blunt tails and short bodies that resemble a pipe, thus the common name. Both are blackish, but have different colour patches.

Sunbeam snake

Xenopeltis unicolor, the single species of the family Xenopeltidae, is also a burrowing species which grows over a metre in length. A nocturnal snake, it is common both in and outside the forest. The scales are smooth and highly polished, and it is uniformly brownish to grey. Its local name is derived from the display of various colours when its iridescent body is struck by the sun. A very docile species, it has been a favoured pet of snake lovers.

Water snakes

The family Acrochordidae contains two species. Both live entirely in the water, and feed on fish and eels. The elephant trunk snake (*Acrochordus javanicus*), with a very thickset body about 2 metres long, is found in inland freshwater rivers while the file snake (*Chersydrus granulatus*), a slender snake about 1 metre long, is confined to river mouths and along the coast. Formerly abundant, the elephant trunk snake is now rare because of pollution and also because it is highly prized for its skin.

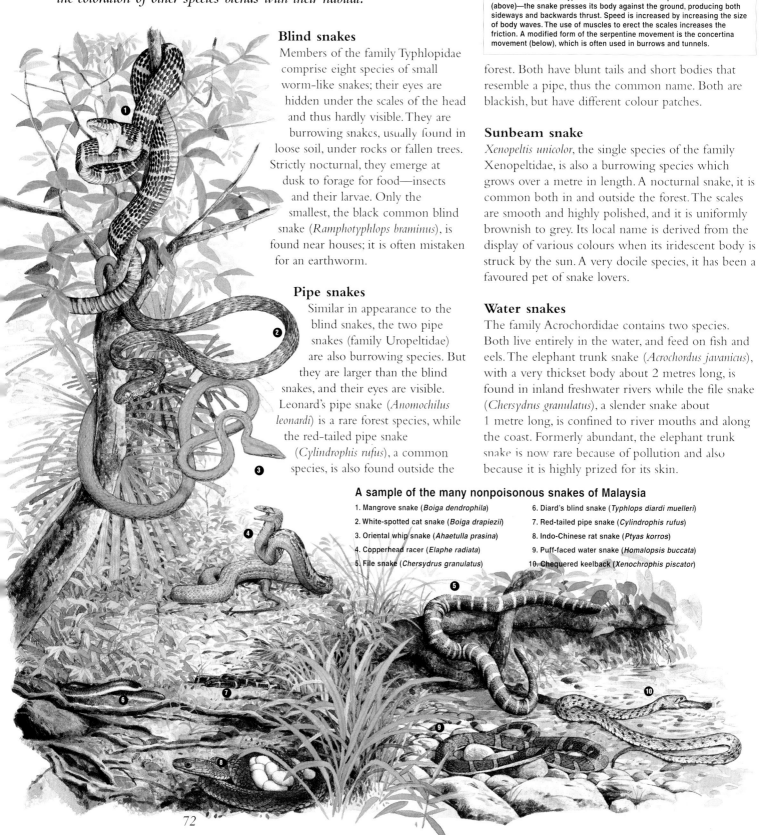

A sample of the many nonpoisonous snakes of Malaysia

1. Mangrove snake (*Boiga dendrophila*)
2. White-spotted cat snake (*Boiga drapiezii*)
3. Oriental whip snake (*Ahaetulla prasina*)
4. Copperhead racer (*Elaphe radiata*)
5. File snake (*Chersydrus granulatus*)
6. Diard's blind snake (*Typhlops diardi muelleri*)
7. Red-tailed pipe snake (*Cylindrophis rufus*)
8. Indo-Chinese rat snake (*Ptyas korros*)
9. Puff-faced water snake (*Homalopsis buccata*)
10. Chequered keelback (*Xenochrophis piscator*)

Malaysian pythons, the constrictors

Although nonpoisonous, the two Malaysian pythons, belonging to the family Boidae, differ from other snakes in constricting their victims before devouring them. Cases of pythons constricting and swallowing humans are very rare. However, in a recent case in Malaysia a reticulated python (*Python reticulatus*) wrapped itself around a man, killing him, but the snake could not work its head over the man's shoulders to swallow his body.

The reticulated python is the second longest snake in the world; 5–7 metres is common, but it can exceed 9 metres, and weigh more than 100 kilograms. Its girth may measure 60 centimetres. The colour is light brown, darker in older and larger snakes, with a complicated pattern of black reticulating lines.

The short-tailed python (*Python curtus*) is a heavily built snake; adults measure less than 3 metres long. Colour is variable—shades of brown to even bright red; hence it is also known as the blood python.

The female reticulated python lays 30–70 oblong white eggs which measure 3.5 x 2 centimetres. It takes 2–5 days to lay the eggs and 90–103 days for them to hatch; the female incubates the eggs until they hatch. The average hatchling length is 60–75 centimetres.

Pythons are generally nocturnal, and are lethargic and slow-moving snakes. They move in rectilinear progression; the body progresses in a straight line like a millipede with waves of motion of the ribs in quick succession. The reticulated python can also climb in search of prey and often hides in trees; the short-tailed

python, in contrast, is strictly terrestrial. Both python species are at home in water, especially the short-tailed python, which lies partially submerged while awaiting prey. Pythons also lie in wait near jungle trails to capture passing animals.

Pythons feed indiscriminately, but prefer mammals such as rats, mouse deer, civets, wild boar and domesticated animals. They kill their prey by constricting it. During the slow swallowing process, the wind pipe expands to assist breathing, and if the python is disturbed during this process the prey is regurgitated. After a heavy meal, the snake coils itself in a safe place to slowly digest its food. A meal of a wild boar of 20–25 kilograms takes at least a full week to completely digest.

Like other wild animal meat, python flesh is believed to have medicinal properties. Mixed with rice wine and eaten raw, the gall bladder is said to be a blood purifying agent, and also an aphrodisiac. Python skins are highly prized. The export trade in python skins is worth millions of ringgit annually. Skins are tanned for the manufacture of shoes, handbags and other fashion items. Due to the demand for

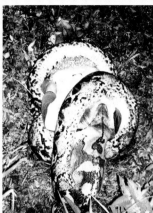

A report in the *New Straits Times* on the reticulated python (*Python reticulatus*) which killed a rubber tapper in a Malaysian plantation in 1995 by strangling him. However, the snake was unable to swallow the man, as it usually swallows its prey, because the victim's shoulders were too wide for the snake's jaws.

both meat and skin, both python species are threatened, even though they are protected by law. Though pythons are bred in other countries for the pet trade, this is not done in Malaysia.

The short-tailed python (*Python curtus*), a much stouter snake than the reticulated python, is also known as the blood python as some specimens have blood-like colouring.

Eggs of a reticulated python (*Python reticulatus*) hatching soon after the mother snake was captured because it nearly attacked a child.

A python's skull shown at rest and with jaw agape ready to swallow prey. The bones connecting the upper and lower portions of the skull function like a hinge, allowing the python to stretch its mouth to many times its normal size. When open, the sides of the lower jaw come apart at the front.

A cross section of a python's stomach before and after it has swallowed prey: The rib bones are only connected to the backbone, unlike in humans, where the rib bones are connected, forming a complete circle. Thus the rib bones of the python can stretch to accommodate a large animal.

Colubrid snakes

Most of the nonpoisonous snakes in Malaysia (89.5 per cent or 129 species) belong to the Colubridae family, which is divided into six subfamilies, from widely varying habitats—ground, trees and water.

Of the nine species of slug snakes all are ground-dwelling snakes except the arboreal blunt-headed tree snake (*Aplopeltura boa*). All are small drab-coloured forest snakes 30–80 centimetres long. The ground species all feed on snails, slugs and insect larvae, while the blunt-headed tree snake feeds on small lizards and geckos.

The 24 reed snakes are dark-coloured ground-dwelling forest species, 30–50 centimetres long, most uncommon; the most common is the variable reed snake (*Calamaria lumbrocoidea*). All species feed on insects and their larvae; some are cannibalistic, while others also feed on small skinks and lizards. The keeled reed snake (*Xenodermus javanicus*), a very rare grey forest species with a maximum length of 65 centimetres, is in a subfamily of its own.

In some of the 18 species of keelback snakes, whose name comes from the keeled dorsal scales, the saliva may be more or less venomous. Mostly ground dwellers, though some are partially aquatic, the keelbacks are 80–100 centimetres long and prey on lizards, frogs, small rats and small snakes.

Though aquatic, the 14 freshwater snakes can move about on land. Stout and about 50 centimetres long, all species feed on fish and crustaceans, including molluscs.

Most of the 63 species of tree and ground snakes are very colourful, and range from 30 centimetres to 3 metres long. Those with spectacular colours and commonly sighted include the paradise tree snake (*Chrysopelea paradisi*), green whip snake (*Ahaetulla prasina*), mangrove snake (*Boiga dendrophila*), red-tailed racer (*Gonyosoma oxycephalum*) and cave snake (*Elaphe taeniura*). The majority of the snakes in this subfamily are rat eaters, and a number of species are found in oil palm plantations, sugar cane plantations and paddy fields where rats are plentiful. Forest rats are also kept in check by the snake population.

A traditional carved door from a Sarawak longhouse, featuring snakes, lizards and frogs.

Frogs and toads

Malaysia has more than 165 species of frogs and toads, which live mainly in the rainforest and in wetland areas. There are a number of endemic species, some unique to Peninsular Malaysia, others to Sabah and Sarawak. Among the unusual frogs in Malaysia are three species of tree frogs which can 'parachute' from the tree canopy to the forest floor, and another species whose eggs develop directly into froglets.

The Malayan corrugated frog (*Rana laticeps*), aquatic in habit, has a call resembling the sound of a bottle being filled with a stream of water.

The red-eyed ground frog (*Leptobrachium hendricksoni*), which has the largest tadpoles in the Malaysian region.

The noisy froglet (*Microhyla butleri*), a tiny frog which is commonly found in swampy roadside spots in Malaysia.

The Malayan jewel frog (*Rana signata*) is found only at the confluence of forest rivers.

Moyang katak dan kala

A Mah Meri carving of a powerful frog holding aloft an evil scorpion, an illustration of a traditional legend from the time when animals had the ability to turn themselves into humans.

To avenge the murder of his brother, a scorpion-turned-human strangled his girlfriend, who turned into a frog. With the frog and his brother's body beside a fire, he cut leaves of the kelemoyang plant into seven pieces. His murdered brother was brought back to life, but the frog remained unchanged. The 'wak-wak' ('father' in the Mah Meri language) call of the frogs still reminds the people of the time when frogs were human.

Frogs and toads

Frogs (*katak*) and toads (*kodok*) belong to the class Amphibia as they live in two separate realms—land and water. As they are tailless in adulthood, they are placed in the order Anura, which means 'without tail'. In tropical countries, the usual differences (roughness of skin and length of hind legs) between frogs and toads are not distinct. The number of frog and toad species in the world is estimated to exceed 2,000. They can be found on every continent; each main zoogeographical region has its own respective fauna. Large islands, such as Borneo and New Guinea in the tropics, also exhibit specific frog and toad fauna of their own.

Malaysian species

The frog and toad fauna of Peninsular Malaysia and of Sabah and Sarawak both have their own endemics because of the different orientation of their general fauna. Peninsular Malaysia has a strong relationship in species composition with the adjacent island of Sumatra and the landmass to the north known as the Indochinese subregion. The frog and toad fauna of Sabah and Sarawak have a strong similarity to Kalimantan of Indonesia, which shares the island of Borneo. The endemicity is much higher in these states due to the large land area of Borneo, which allows for a more extensive evolution of these animals. Sabah and Sarawak share a portion of their frog and toad fauna with Peninsular Malaysia and also other islands in the Malaysian subregion.

The tailed tree frog (*Rhacophorus promianus*), an endemic of the foothill and montane forests of Peninsular Malaysia.

Habitat

Most frogs and toads live in the rainforest and wetland areas. Only a few species live near human habitation in urban and suburban areas. *Bufo melanostictus*, *Kaloula pulchra*, *Microhyla heymonsi*, *Polypedates leucomystax* and *Rana limnocharis* are the most common of these.

The only frog species known to occasionally venture into the sea is *Rana cancrivora*, which lives in mangrove areas in Southeast Asia and is also found in inland swampy areas. This species has a kidney with a greater than normal capacity to secrete salt. However, it is not able to withstand prolonged immersion in sea water. Its tadpoles are not known to be able to live in sea water, and the species needs bodies of freshwater for breeding.

Among the frog and toad species in the montane mossy forests of Malaysia are *Megophrys aceras*, *M. longipes*, *Metaphrynella pollicaris*, *Philautus aurifasciatus*, *P. vermiculatus*, and *Rhacophorus bipunctatus*. *Metaphrynella pollicaris* lives and breeds in a tree hole in the forest canopy, and its 'poh-poh' call can often be heard in the evenings. The horn toads *Megophrys aceras* and *M. longipes* lay their eggs in ground pitchers of *Nepenthes* plants.

In the complex structure of the lowland forest, frogs and toads are spread in vertical distribution from the top of the canopy down to the leaf litter on the forest floor. They tend to aggregate along rivers and streams, and in seepage areas and pools and puddles where they breed and replenish their body fluids with water. The tadpoles of species adapted to living in rapids and waterfalls, such as genera *Amolops* and *Staurois*, have ventral suckers to attach themselves to rocks so they are not swept away by the swift current.

Defence

On the forest floor is the master of camouflage of the frog world, a horn toad, *Megophrys monticola nasuta*, which resembles one of the leaves in which it is hiding when viewed from above. The skin flaps over the eyes are used to conceal the eyes from potential predators when the toad remains motionless. This species also provides a warning of imminent rain. Its metallic 'kuang' call in the forest is a sure sign that rain will soon arrive. The tympanum of the ear of this species is highly sensitive to changes in atmospheric barometric pressure; thus its call is induced by a drop in atmospheric pressure resulting from the formation of rain.

The microhylid frogs living among the leaf litter of the forest floor often burrow into loose soil for protection from predators. Individuals have been observed to have their favourite spots, although these are not permanent.

Large and small

The largest toad in Malaysia—found in Sabah and Sarawak—is *Bufo juxtasper*, which can weigh more than 600 grams. In Peninsular Malaysia, the largest toad is the giant Malayan toad (*Bufo asper*), which is often seen on garden lawns of houses near streams, and in streams and rivers in lowland forest. It is also the most common toad found in caves in Malaysia.

The largest frog in Malaysia is the Malayan giant frog (*Rana blythi*), reported to grow to a length of 26 centimetres from snout to vent, and to weigh over 700 grams. A specimen from Sumatra has been recorded in the *Guinness Book of Records* as among the largest frogs in the world. Large specimens have been reported to eat birds, rats, snakes, centipedes and scorpions.

At the other end of the scale, the smallest members of the frogs and toads are among the genus *Pelophryne* in the family Bufonidae and the genus *Microhyla* in the family Microhylidae. An adult of these species weighs as little as 10 grams, and can comfortably sit on a person's thumbnail.

Poisonous toads

In Malaysia, there is no equivalent of the poison arrow frogs of South America. However, the toads in the family Bufonidae all have poison secreted by the glands on their skin. This bitter and pungent tasting poison can irritate the mucus membranes of the

Gourmet delicacy

The frogs commonly caught and eaten in Malaysia are *Rana blythi* (right), *R. ingeri*, *R. cancrivora* and *R. limnocharis*. Frogs reared in Malaysia for the table are American species introduced through Taiwanese frog farms. The common American bullfrog (*Rana catesbiana*) has been successfully reared on formulated pellets and is currently marketed, mainly to Chinese restaurants which sell frog meat as part of their gourmet cuisine.

Flying or parachuting frogs?

Wallace's flying frog (*Rhacophorus nigropalmatus*) and its sister species, *Rh. pardalis* (below) and *Rh. reinwardti*, are tree frogs which forage for food high up in the canopy of Malaysian rainforests. To escape predators, they launch themselves from the canopy and parachute down to the understorey trees or the forest floor. The parachuting of these frogs is often mistaken for flying—which would require the frog to travel through the air at an angle of less than 45 degrees with some means of powered propulsion. They have the ability to parachute because of the extended webbing on their hands and feet as well as other skin extensions on their arms and legs which match the air-like parts of a sail to reduce the speed of descent.

Wallace's flying frog (above) is named after Alfred Russel Wallace, who discovered the frog in Borneo, describing it thus in his *The Malay Archipelago*: *I found the toes very long and fully webbed to their very extremity, so that when expanded they offered a surface much larger than the body. The fore legs were also bordered by a membrane, and the body was capable of considerable inflation. The back and limbs were of a very deep shining green colour, the under surface and the inner toes yellow, while the webs were black, rayed with yellow. The body was about four inches long, while the webs of each hind foot, when fully expanded, covered a surface of four square inches, and the webs of all the feet together about twelve square inches.*

mouth or other moist areas of a person. Such poison has been reported to be used in local black magic potions to cause stomach upsets or disorders, and is also used in Chinese medicine for treating heart problems.

Frog wars

Frogs and toads rarely attract human attention. The report of frog wars, which are believed to be an indication of the coming of war among humans, is rare. An investigation of one such report revealed the occurrence of mass breeding of common species in newly formed ponds after heavy rain following a prolonged drought. The wrestling among the frogs for breeding territories had given the impression to the human observers that the frogs were at war with each other. Imaginative individual observers have often added spice to their story by reporting the use of twigs as swords and leaves as shields by the frogs, thus making their tall story much more convincing.

Conservation

The frog and toad population in many developed areas of Malaysia, as in the rest of the world, is declining. Forest species seem to be the main victims of habitat destruction, while riverine species tend to fall victim to water pollution. Tadpoles are highly vulnerable to chemical pollutants in the water, while adults can also be killed by such pollutants in the air and soil. However, common commensal species such as *Bufo melanostictus*, *Polypedates leucomystax* and *Rana limnocharis* seem able to survive the threats to which other more sensitive species are succumbing.

Frogs and pitcher plants

Pitcher plants (*Nepenthes* spp.) have tubular, hollow leaves modified as pitchers, which are lined with downward-pointing hairs. Insects which enter a pitcher thus cannot escape and fall to the bottom, where they are digested by the juices there. The pitcher plants usually grow on poor soil, and depend on insects for nutrients. However, it is not only insects which find their way into such pitchers.

The long-legged horned toads (*Megophrys longipes* (top) and *M. aceras*) lay their eggs in ground pitchers. The tadpoles are able to resist the digestive power of the fluid in the pitchers. *Philautus aurifasciatus* chooses old pitchers in which to lay small clutches of its large eggs (inset), which develop directly into froglets without a tadpole stage.

In the mangroves along Malaysia's coastline (above), many species of colourful crabs, as well as other fascinating animals such as mudskippers—fish which are able to survive for extended periods out of water—are seen. Freshwater habitats (below) support a diversity of aquatic life, such as fish, crabs, prawns and snails.

AQUATIC ANIMALS

Malaysia's freshwater ecosystems can be divided into two distinct types: lentic (still water) and lotic (running water). They support a great variety of animal life. In the broadest sense, Malaysia's aquatic animals represent all groups of animals, from protozoans to mammals. They can be divided into surface dwellers (such as water skaters and whirligig beetles), floating forms (zooplankton, such as water fleas, rotifers and copepods), swimmers (fish, amphibians, reptiles, birds and mammals), stone and weed dwellers (insects, snails and shrimps), and burrowers or bottom dwellers (annelid worms and clams). However, many of these animals (for example, mammals and reptiles) are covered in other parts of this volume and this section concentrates on fish, crabs, prawns and snails.

While some of the freshwater fishes of Malaysia are highly prized by the restaurant trade for their delicious flavour, others, such as the golden dragon (*kelesa*) (*Scleropages formosus*), fetch high prices in the aquarium fish trade for their beauty. Although not of economic importance in Malaysia, another family of fishes, the mudskippers, are endlessly fascinating to visitors to the mud flats, with their ability to travel across land and to survive out of water for long periods.

Also very noticeable in the mangroves at low tide are the fiddler, soldier and hermit crabs, especially the very colourful species, and their distinctive burrows. Most striking are the male fiddler crabs with one extra large claw, which they use for signalling females and for fighting to defend their territory. Now confined to only a few islands is the very large hermit crab, the coconut or robber crab (*Birgus latro*).

Crabs of a different type inhabit the rivers of Malaysia. More than 90 species have already been identified, and others are still awaiting discovery. Some members of this very diverse group must return to the sea to breed. Others complete their life cycle entirely in fresh water. Yet others have adapted so well that they have become completely terrestrial; some live in caves.

Similarly, some of Malaysia's freshwater prawns must return to the sea to complete their life cycle, while others spend their entire life in fresh water. The large *Macrobrachium* prawns with very long pincers are considered a gourmet delicacy, and are being cultivated on a large scale in aquaculture projects for export as well as to fulfil the insatiable demand from local restaurants.

Freshwater snails are also marine creatures which have adapted to life in fresh water. Although some freshwater snails act as biological control agents, others are seen in a less favourable light. The golden apple snail (*Pomacea canaliculata*) is a pest in the paddy fields, while a number of species are of public health importance as the hosts of parasites from which people can contract diseases.

A selection of the diverse animals to be found in two aquatic environments in Malaysia: mangroves and a freshwater habitat.

MANGROVE
1. Mudskipper (*Boleophthalmus boddaerti*)
2. Mudskipper (*Periophthalmus chrysospilos*)
3. Sand-bubbler crab (*Scopimera* sp.)
4. Soldier crab (*Dotilla myctiroides*)
5. Mud snail (*Cerithidea* sp.)
6. Fiddler crab (*Uca dussumieri*)
7. Ghost crab (*Ocypode ceratophthalma*)
8. Fiddler crab (*Uca rosea*)
9. Fiddler crab (*Uca triangularis*)
10. Fiddler crab (*Uca rosea*)
11. Hermit crab in shell of *Terebratia* sp. mangrove snail

FRESH WATER
12. Golden dragon (*Scleropages formosus*)
13. Snakehead (*Channa micropeltes*)
14. *Gyraulus* sp. snail
15. *Melanoides* sp. snail
16. Carp (*Rasbora* sp.)
17. Puffer fish (*Tetraodon* sp.)
18. Walking catfish (*Clarias batrachus*)
19. *Pila* sp. snail
20. *Macrobrachium* sp. prawn
21. Two-spot gourami (*Trichogaster trichopteris*)
22. *Potamos* sp. crab
23. *Gyraulus* sp. snail
24. *Caridina* sp. prawn
25. *Geosesarma* sp. crab
26. T-barb (*Puntius lateristriga*)
27. Climbing perch (*Anabas testudineus*)

A bright red male fiddler crab (*Uca rosea*), which uses its large claw to attract a mate.

Freshwater ecosystems

Fresh water is essential to the maintenance of both aquatic and terrestrial life, including man, although only about 1 per cent of water on the earth is fresh water. It differs from sea water in its salinity value, which is the number of grams of salts of sodium, potassium, calcium, magnesium, chloride, sulphide, carbonate and bromide dissolved in 1 kilogram of water. Freshwater ecosystems are divided into two systems, lentic (still water) and lotic (running water).

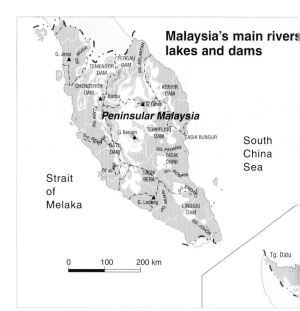

Malaysia's main rivers, lakes and dams

Peninsular Malaysia

Strait of Melaka

South China Sea

0   100   200 km

Tg. Datu

A common kingfisher (*Alcedo atthis*), which relies on the freshwater ecosystem for its fish catch.

Some fishes of the Malaysian freshwater ecosystem

1. *Puntius lateristriga*, common in undisturbed streams flowing through lowland forests.

2. *Rasbora elegans*, which feeds mainly on insects which fall into torrential streams, its favoured habitat.

3. *Betta pugnax*, which lives in stagnant pools with leaf litter substrate.

Lentic ecosystems

The technical term lentic, which is derived from the Latin *lentus*, meaning 'sluggish', refers to still waters such as lakes, ponds (generally smaller and shallower than lakes), swamps (wet lowlands with mosses, shrubs and large trees) and marshes (treeless wetland areas with grasses, rushes and sedges).

In Malaysia, there are no lakes formed by natural means, such as retreats of continental glaciers, volcanic eruptions and tectonic folding and faulting. Many large bodies of water in Malaysia which can be classified as lakes are reservoirs, formed by the construction of a dam across a river. These include Kenyir Dam in Terengganu and Temengor and Chenderoh Dams in Perak (built for power generation), Linggiu Dam in Johor (built for water supply), and Batu Dam in Batu Caves, Selangor (built for flood mitigation). The ability of these man-made lakes to support life is dependent on their physical structure. The deeper the lake and the higher the water temperature, the less the lake is habitable. Temengor Dam, with a maximum depth of more than 60 metres and surface water temperature exceeding 27 °C, cannot support fish life below a depth of 20 metres. The number of fish starts to decrease tremendously at a depth of 6 metres. High concentrations of toxic gases, particularly hydrogen sulphide, and a lack of dissolved oxygen, are two main reasons for the reduction of fish life in deep waters.

Swamps have traditionally been considered unproductive areas, and many have been drained to make way for plantations and other development. However, studies of the Malaysian freshwater ecosystem have shown the importance of the swamp for fish production, flood control and production of forest products. Peatswamps, such as those in north Selangor, have a substrate of peat, comprising

undecayed plant material. The peat releases into the water a yellowish, astringent substance called tannin and other organic acids, giving it a brownish tint. Because of its high acidity, the peatswamp is considered the harshest aquatic ecosystem in the world. Fish species and other aquatic life not adapted to live in the peatswamp habitat succumb to the toxic material. Although the habitat is uncommon in Malaysia, many fishes have become so dependent on the habitat that they do not reproduce without the black waters.

More common in Malaysia are freshwater swamps such as Tasik Bera, Tasik Chini and Tasik Bungur. These swamps are characterized by floating vegetation such as *Utricularia* (bladderwort) and *Nymphaea* (water lily) in the surface water. At the fringe of the swamp, *Lepironia* reeds (*purun danau*), *Pandanus* (*mengkuang*) clumps and *Eugenia* (*kelat*) swamp forest stands are common vegetation.

Lotic ecosystems

Flowing waters such as rivers and streams are grouped under the term 'lotic' ecosystems, from the Latin *lotus*, meaning 'to wash'.

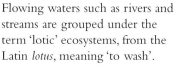

A freshwater ecosystem

A representation of various parts of the Malaysian freshwater ecosystem, following a river from a highland waterfall, as a fast-moving mountain stream, slowing as it reaches the lowlands, before finally emptying into a water lily-covered freshwater swamp. Before reaching the swamp, water from the river is diverted into a man-made irrigation canal for the paddy fields.

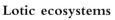

Rain supplies the water in these ecosystems.

The lotic ecosystem in the upper reaches consists of mountain streams passing through steep gradients. The water flow is very fast and turbulent. Undisturbed streams in the mountains are characterized by crystal clear water, low temperature and high dissolved oxygen content, with little seasonal variation. Aquatic organisms living in this ecosystem are specialized groups able to tolerate only a small range of environmental changes. They are bottom dwellers, some equipped with a hold-fast organ to prevent them being washed away downstream. A small catfish, locally known as *ikan keli depu* (*Glyptothorax major*), with a suction organ on its belly, attaches itself to rocks to prevent itself being washed downstream while browsing for food. The caddisfly, an aquatic insect of the order Trichoptera, attaches its nest to the bottom substrate of the stream so it can filter food in the torrential stream.

Disturbances near the streams that affect dissolved oxygen concentration, water temperature, clarity or suspended load have a profound effect on the organisms. The montane aquatic dwellers are the first to become locally extinct with only slight changes to these habitats. For instance, more than 50 per cent of the fish community in the upper reaches of the Gombak River have disappeared due to the increased sediment load. Fishes such as the montane catfish (*Hemibagrus gracilis* and *Glyptothorax platypogonoides*) and the pipefish (*Doryichthys deokhatoides*), previously common in the river, have become extinct.

Lowland rivers consist of middle and lower stream courses. The middle stream course typically has a greater incline gradient and greater velocity than the lower course. The river bed has coarser materials, with mud and lighter sediments in pools and sidewaters. As a stream approaches its mouth, the bottom is composed of loose mud, silt and organic detritus, and the shores generally bordered by swamps. Because of the low gradient and the nature of the soil, lowland rivers in Malaysia have high concentrations of suspended particles.

The absence of canopy cover results in a higher water temperature than in the montane streams. Coupled with low dissolved oxygen concentration, the habitat is suitable for organisms able to tolerate a wider range of environmental changes. Typical organisms are mud dwellers or free swimmers, feeding mainly on detritus. Some are scavengers, feeding on dead organisms. In clean rivers, such as the Rompin River in Pahang, colonies of aquatic flies, mayflies, snails, dragonflies, and shrimps are common inhabitants. In polluted rivers, such as the Klang River in Selangor, the substrate is dominated by bloodworms; animal diversity is extremely low.

Creatures of the freshwater ecosystem

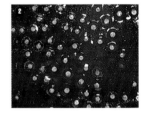

Among the myriad small creatures of the freshwater ecosystem are:

1. Giant Malayan toad (*Bufo asper*)
2. Amphibian eggs
3. Tadpole of *Polypedates leucomystax*
4. Stonefly (Plecoptera) nymph
5. *Parapoynx bilinealis*, an aquatic moth
6. Pond skater (*Tenagogonus cillatus*)

Carps and catfishes

There are more than 1,500 species of carps in the world, forming the family Cyprinidae. Of these, 100 species occur naturally in Malaysia, including a few endemics, while other species have been introduced. Malaysia also has 50 species of catfishes, which represent only a small fraction of the 2,000 widely diverse species of catfishes (divided among about 34 families) present worldwide.

A Fisheries Department freshwater fish research station at Kenyir Dam, Terengganu, which also serves as a tourist attraction. Research is essential for the expanding freshwater fish industry, which supplies ever increasing quantities of fish to both local and export markets.

INTRODUCED CARP SPECIES

COMMON NAME	SCIENTIFIC NAME
Bighead carp	*Hypophthalmichthys molitrix*
Bighead carp	*Hypophthalmichthys nobilis*
Common carp	*Cyprinus carpio*
Grass carp	*Ctenopharyngodon idella*
Indian carp	*Cirrhinus mrigala*
Indian carp	*Catla catla*
Javanese carp	*Puntius gonionotus*

NATIVE CARP SPECIES PRIZED AS FOOD

MALAY NAME	SCIENTIFIC NAME
Ikan jelawat	*Leptobarbus hoevenii*
Ikan kelabau	*Osteochilus melanopleurus*
Ikan kelah	*Tor tambra*
Ikan krai kunyit	*Puntius pierrei*
Ikan lampam sungai	*Puntius schwanenfeldii*
ikan seluang kunyit	*Rasbora tornieri*
Ikan temegalan	*Puntioplites bulu*
Ikan tengas	*Neolissochilus soroides*

Similarities

Carps and catfishes share a common feature in the possession of a Weberian apparatus, a connection of alternating ligaments and small bones between the gas bladder and the inner ear. The structure, which is derived from the first four vertebrae, facilitates sound reception, allowing carps and catfishes to detect a wide range of environmental changes in the aquatic ecosystem.

Carps

With 1,500 species of carps, the family Cyprinidae is undoubtedly the most diverse group not only among fishes, but among all vertebrates. They are found in most countries of the world. Of the 100 species of carps naturally present in Malaysia, 70 species are known to inhabit the Pahang River basin, the largest number known in a single river in this country. A few carp species are endemic, such as *Osteochilus ingeri*, which is endemic to Sabah, and *O. sarawakensis*, which is endemic to Sarawak. Some carps are very tiny and colourful, while others can weigh as much as 20 kilograms.

Carps are characterized by a moderately compressed body with lower and upper profiles equally convex, covered with scales. The mouth is more or less protrusile, with thin lips, and is always toothless. Teeth are present only on the pharyngeal bone, in 1–3 rows, with not more than nine teeth in any one row. There are usually one or two pairs of barbels (whiskers), sensory structures for locating food; however, a few species do not possess any barbels. The tail fin is deeply to slightly forked with equal lobes. A spine-like ray is present in the dorsal fin of many carps. Species feeding on planktonic organisms may have a well-developed gill raker situated on the front margin of the first gill arch.

Many non-native species have been introduced into Malaysia, some as highly prized food fish, others for the aquarium trade. The most numerous is the common carp (*Cyprinus carpio*), locally known as *ikan lee koh* or *ikan kap*, which is reared in ponds and mining pools as well as in reservoirs. The species has been introduced throughout much of the world from its native land, China. The common carp is a delicious fish if prepared properly, and for this reason it is considered a delicacy in the Oriental region as well as in some European countries. However, in North America the common carp is considered a pest, apparently because while rooting about for food, it often roils the water, increasing turbidity and adversely affecting many plants and other aquatic animals, particularly native fishes of America such as the highly esteemed trout.

Many carp species are of importance as food. The highly prized *ikan kelah* (*Tor tambra*) is one of the most sought after fishes by restaurants specializing in freshwater fish cuisine. Some members of the carps of the genus *Rasbora*, locally known as *ikan seluang*, are also an important source of protein for people staying along the river. For instance, in the Kelantan River, *ikan seluang kunyit* (*Rasbora tornieri*) is baited by spreading grated coconut on the surface of the water. It is then caught with a cast net. The fish is best prepared by adding a little salt and tamarind, wrapping with a banana leaf and baking for a few minutes. This dish is locally known as *ikan pais asam*, and is unquestionably delicious.

Carps

A 10-kilogram *ikan temoleh* (*Probarbus jullieni*) caught in Taman Negara.

Ikan temegalan (*Puntioplites bulu*), a highly esteemed food fish which has adapted to living in impoundments such as the Temengor and Kenyir Dams.

Ikan mata merah (*Cyclocheilichthys armatus*) is found in groups in shallow sections of medium-sized rivers flowing through forested areas.

Ikan tengas (*Neolissochilus soroides*), a popular fish found in large montane streams with a rocky substrate.

Ikan sebarau (*Hampala macrolepidota*), an important game fish which can reach 70 centimetres in total length.

Ikan baung pucuk pisang (*Leiocassis micropogon*), which is found in medium-sized rivers flowing through forested areas.

Ikan tapah bemban (*Ompok bimaculatus*), a small carnivorous fish which feeds on tiny fish and other aquatic animals in small lowland forest streams.

Ikan tapah (*Wallago leerii*). Its mouth is so wide it is believed to be able to swallow animals as large as the long-tailed macaque.

Ikan baung kunyit (*Hemibagrus nemurus*), which is considered to be one of the most delicious catfishes in Malaysia.

The barbels near the mouth of *Clarias teijsmanni*, somewhat resembling the whiskers of a cat, are used to find food and detect environmental changes.

Mystus bimaculatus, a small catfish found in blackwater streams and peat forests, emits a chirping sound if lifted out of the water.

Ikan tapah (*Wallago leerii*), Malaysia's largest freshwater fish, caught in Sungai Tembeling, Pahang.

Catfishes

With more than 2,000 species in about 34 families, the catfishes are probably the most diverse group in the world, at both species and family levels. Catfishes have an elongated body, smooth, scaleless skin, and 1–4 pairs of long and slender barbels. The name 'catfish' derives from the presence of these barbels near the mouth, somewhat resembling the whiskers of a cat. The fins are supported by soft rays, except for a stout, sharp spine in front of the dorsal fin and each of the pectoral fins which can pierce painfully. Although there is no toxin released by the spines, swelling may result if the wound is covered with mucus produced by the skin of the catfish, which is known to contain a toxic chemical. Traditionally, it was believed that applying the brain of the offending fish to the wound was an effective antidote. A small rayless fatty fin is also present on the midline of the back just ahead of the tail fin in many species of catfishes. Hundreds of tiny teeth are arranged in bands on the roof of the mouth. The eyes are relatively very small, and the mouth cannot be extended.

Most catfishes inhabit freshwater environments, but some are found in estuaries and seas. The freshwater catfishes are the most diverse group in terms of the number of families. Among Malaysia's 50 species of catfishes, some, such as the montane catfish *Glyptothorax major*, are very small, reaching a maximum length of about 5 centimetres. Others, such as *ikan tapah* (*Wallago leerii*) and *ikan kenderap* (*Bagarius yarrelli*), can be huge, reaching more than 150 centimetres in length.

Since catfishes have no bone in their muscle and the flesh is very tasty, they are excellent food fish. An African catfish of the genus *Clarias* is cultured extensively in Malaysia by small-scale fishermen. It is usually sold live in fish markets.

Some catfishes are very common, while others are very rare. The catfish of the genus *Hemibagrus*, locally known as *ikan baung*, and the catfish of the genus *Kryptopterus*, popularly called *ikan selais*, are the most diverse group in Malaysia. *Ikan baung kunyit* (*Hemibagrus nemurus*) is a very tasty fish and may fetch a high price. On the other hand, some *ikan selais* are very small and so beautiful that they are popular as ornamental fishes.

Like their namesakes the cats, catfishes are most active at night. During the day, they often hide under submerged logs, in undercut banks, crevices and cavities, or they remain quietly in the deeper parts of pools. At night, catfishes hunt for food, using their external taste sensors, which are most abundant on the whiskers and lips. Undoubtedly, to catfishes taste and touch play a more important role than sight in feeding and general orientation to the environment. Although they are generally active at night, a sudden sharp rise in the level of the stream or pool may also cause catfishes to go on a feeding spree and forage actively at all hours of the day or night. Therefore, fishermen who spin their reels at dusk or after heavy rains may be able to catch large numbers of catfishes.

A few catfishes have an accessory breathing organ called a labyrinth or a gill tree in the upper part of each brachial cavity. This allows them to live in oxygen-poor water, particularly in oxidation ponds, paddy fields and swamps. The genus *Clarias*, especially *Clarias batrachus*, commonly known as *ikan keli*, is the most tolerant to such an environment. These species can leave the water and crawl about on land, and for this reason they are called 'walking catfish'. Members of the genus *Clarias* are so dependent on air that they will drown if deprived of the opportunity to breathe in fresh air.

Storage baskets placed in water for keeping fish alive.

Three traditional fish traps used for catching freshwater fish in Malaysian rivers.

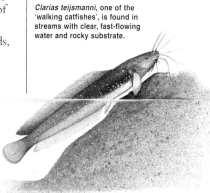

Clarias teijsmanni, one of the 'walking catfishes', is found in streams with clear, fast-flowing water and rocky substrate.

81

Aquarium fishes

The beauty and brilliant colours of Malaysian freshwater fishes, coupled with their good temperament, have made them very popular as aquarium fishes all over the world. The golden dragon fish (Scleropages formosus) is probably the most decorative aquarium fish of Malaysia. The belief that it will bring prosperity to the owner has resulted in the species becoming one of the most expensive freshwater aquarium fishes in the world. Other popular aquarium fishes include the gouramis, snakeheads, rasborine and puntiid carps, catfishes and puffer fishes.

The beautiful golden dragon (*Scleropages formosus*), probably the world's most expensive aquarium fish, which is thought by many people to bring good luck to its owner.

Tanks of aquarium fishes in one of the numerous shops catering for the demand for these fishes from Malaysians who have aquariums in their homes and offices.

Golden dragon

Probably the most expensive of all aquarium fishes in Malaysia is the golden dragon fish (*ikan kelesa*) (*Scleropages formosus*) belonging to the family Osteoglossidae. Members of this family are found in South America, Africa and Australia, as well as in Southeast Asia. The *ikan kelesa* is extremely beautiful, and is believed to bring good luck to the owner. As a result, a 50-centimetre specimen can easily fetch a price of more than RM5,000. Due to high demand by aquarists and also because of habitat degradation, the fish has become very rare. Although it has been listed as an endangered species by the International Commission on Endangered Animals, there is no regulation on capture and possession of the species in Malaysia. However, most aquarium specimens are believed to have been bred in captivity.

Gouramis

Gouramis are common aquarium fishes of Malaysia. They have a complex respiratory chamber, situated above the gill arches, which receives swallowed air bubbles and extracts oxygen from them. Gouramis are so dependent on air that they would drown if deprived of aerial breathing. Since gouramis can breathe atmospheric oxygen, they are very hardy fishes which can survive in an aquarium with low concentrations of dissolved oxygen, and therefore physical aeration for the aquarium is not essential.

Parental care is very obvious among gouramis. The fighting fishes (*Betta* spp.) build a foam nest from air bubbles surrounded by a hardened secretion of the mouth, aquatic plants and algae, which are woven into the nest. Eggs usually float up into the nest through the buoyancy of their own contained oil droplets. Many gourami species are mouth brooders—the incubation of eggs takes place in the mouth, normally of the male parent. But in the chocolate gourami (*Sphaerichthys osphromenoides*), the females are mouth brooders. The male helps with the eggs, usually laid on the bottom of an aquarium, by spitting them over to the female before they are incubated in the mouth. In contrast, some species simply lay floating eggs in the open water.

Snakeheads

The snakeheads (*haruan*) (*Channa* spp.) are carnivorous fishes which feed on live organisms, particularly other small fishes. An accessory respiratory organ in the form of a simple diverticulum from the gill chamber allows supplementary air breathing and so they can exist in an aquatic environment with low concentrations of dissolved oxygen. Thus, an aquarium stocked with snakeheads does not need physical aeration.

Most species build bubble nests in still water. The fertilized eggs are floated up into the nest and are guarded by one or both parents until they hatch and the fries reach about 4 centimetres in length. Some species, such as *ikan toman* (*Channa micropeltes*), are so fierce in guarding the nest that they attack intruders. There are about eight species of snakeheads in Malaysia. Differences in size, habitat preference, food and habits enable them to flourish in a great variety of waters, both flowing and still.

Carps

Carps of the genus *Rasbora* are some of the most beautiful and colourful aquarium fishes. Many *Rasbora* are small and slender, and inhabit freshwater as well as blackwater swamps, which are mostly slow flowing and overgrown with aquatic macrophytes. A few *Rasbora* species occupy the open water of a stream or river.

The little barbs, another group of carps of the genus *Puntius*, are also popular as aquarium fishes. Some species can exceed 15 centimetres in length, but the great majority of them are small. Puntiid carps are easily contented and peaceful, and are therefore suited to community life with other species of peaceful fishes.

Puffer fishes

Puffer fishes (*ikan buntal*), a group of fish with a globular body, are mostly marine inhabitants. However, Malaysia has three freshwater species, *ikan buntal sungai* (*Tetraodon leiurus*), *ikan buntal gantang* (*Tetraodon palembangensis*) and *ikan buntal selasih* (*Chonerhinos nefastus*).

All puffer fishes possess tetrodotoxin, a toxin which can be fatal to humans. Although the toxin is concentrated in the gonad, skin, liver and intestine, small amounts are also found in muscles. Despite the danger, puffer fishes are a popular Japanese gourmet delicacy, known as *fugu*. It is prepared by specially licensed cooks to minimize the risk of poisoning, though some believe that the risk involved in eating fugu enhances the flavour.

Ikan buntal gantang (*Tetraodon palembangensis*), one of the puffer fishes which, when faced by a predator, inflate themselves with air or water to scare away their enemy. It is shown here in deflated and inflated views.

Aquarium fish breeding

Although many aquarium fishes have been successfully bred in captivity, many are taken from the wild. Populations of aquarium fishes in the wild are important because after several generations of captive breeding, the fish lose their desirable forms and are susceptible to various diseases. Occasionally, it is necessary to re-establish the useful traits from the wild stock. Thus, it is imperative to preserve their natural habitats so that the wild gene pool is maintained.

Worldwide trade in aquarium fishes is a multimillion dollar industry, but so far the contribution of ornamental fishes to the fishing industry of Malaysia has been minimal. If developed further, such trade could not only bring in more money from exports, but could make known the beauty of Malaysia's fishes throughout the world.

Front and side views of the puffer fish spirit (*hantu buntal*) of the Jahut, an Orang Asli group. This spirit is believed to bite anyone who enters the water. The bitten part swells and the person will be sick for 2–3 days.

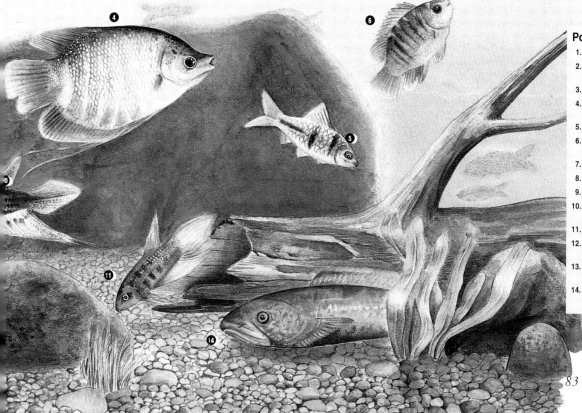

Popular aquarium fishes

1. Queen danio (*Danio regina*)
2. Chocolate gourami (*Sphaerichthys osphromenoides*)
3. Scissortail (*Rasbora trilineata*)
4. Giant gourami (*Osphronemus goramy*)
5. T-barb (*Puntius lateristriga*)
6. Swamp perch (*Pristolepis fasciata*)
7. Leaf fish (*Nandus nebulosus*)
8. Carp (*Crossocheilus oblongus*)
9. Pearl gourami (*Trichogaster leerii*)
10. Croaking gourami (*Trichopsis vittata*)
11. Penang betta (*Betta pugnax*)
12. Banded loach (*Botia hymenophysa*)
13. Freshwater puffer fish (*Tetraodon leiurus*)
14. Montane snakehead (*Channa gachua*)

Mudskippers

Mudskippers are amphibious fish belonging to the family Gobiidae which inhabit muddy estuarine shores from Africa east to the Pacific Ocean, as far south as northern Australia and as far north as southern Japan. There are estimated to be 34 species in the world, of which four are commonly seen on the mud flats of the west coast of Peninsular Malaysia as well as Sabah and Sarawak.

World distribution of mudskippers ranges from the west coast of Africa to the Pacific Ocean

Mudskippers and amphibians

Mudskippers are similar to amphibians in that they spend time both in water and on land, but like other fishes mudskippers have fins instead of land-adapted legs and possess gills throughout their life. Such an existence is unlike that of amphibians, which live as fish-like vertebrates in their early life form but undergo metamorphosis to be terrestrial adults. For their highly peculiar way of life, mudskippers have morphological and physiological adaptations to suit their two vastly different environments. The different physiological systems in the body of a mudskipper have been adapted to tolerate the changes involved in moving from one habitat to the other.

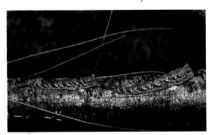

The colouring of this mudskipper provides excellent camouflage, protecting it from predators such as the kingfisher.

Malaysian mudskippers

Locally known as *ikan belacak, ikan belodok* or *ikan tembakul*, the four common Malaysian species of mudskippers, each with a distinctive appearance, are *Boleophthalmus boddaerti, Periophthalmodon schlosseri, Periophthalmus chrysospilos* and *Scartelaos viridis*.

Mudskippers usually live at the fringe of the shore, migrating up and down the intertidal zone according to the tide. The intertidal mud flat coastal region, especially if it has mangrove forest in its vicinity, is a favourable habitat for mudskippers. In Malaysia, mudskippers are commonly found in the soft bottom intertidal areas and estuaries with mangrove wetlands on the west coast of Peninsular

An early account

William T. Hornaday, a noted American zoologist who visited both the Malay Peninsula and Borneo in the 1870s to collect specimens for American museums and who subsequently published *Two Years in the Jungle*, was fascinated by the mudskippers he saw on the coast of Selangor:

The most interesting animals we found on the mud flats were some fishes whose actions were really remarkable. Although apparently stranded there, they seemed to feel perfectly at home, and were jumping round over the mud in every direction with the greatest indifference to their sudden change of element. In reality they were feeding upon the tiny crustaceans left on the bank by the receding tide.

The species is known scientifically as Periophthalmus schlosserii (Pallas, Bl, Schn.) a member of the family Gobiidae, whose expanded ventral fins serve as a foot, the lengthened pectorals as organs of locomotion, while the small gill opening allows the retention of sufficient moisture to sustain the fish for a considerable period on land.

Their burrows were simply mud-holes, going straight down to a depth of three to four feet, large enough in diameter to admit a man's arm easily, and, of course, full of water.

Illustration of a mudskipper from Hornaday's book.

A typical mudskipper

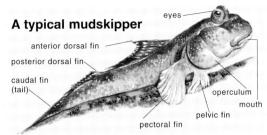

eyes
anterior dorsal fin
posterior dorsal fin
caudal fin (tail)
operculum
mouth
pelvic fin
pectoral fin

The general appearance of a mudskipper is similar to that of any fish. Attached to its aerofoil, elongated body are fins similar to those of fish. However, its closely placed bulging eyes are elevated above the head surface. The position of the eyes is believed to be one of the adjustments to land life made by the mudskipper. The eyes are well adapted for terrestrial and aquatic vision. Each mudskipper species has characteristic colours and patterns (for example, spots or lines) on the body surface.

Malaysia, such as Tanjung Karang, Sementa and Benut, directly facing the Strait of Melaka. They can also be found on the coasts of Sabah and Sarawak. The various species intermingle, that is, they are not limited to definite zones within a common area. In many places, the habitat of the mudskipper is now ecologically threatened by man's activities.

Respiration

The mudskippers use gills and the highly vascularized surface of the buccopharyngeal cavity for respiration. They can tolerate being out of the water for extended periods because of morphological adaptations of the gills. In addition, it is believed that specific parts of the skin (such as the skin of the tail, which is highly vascularized) are involved in dermal respiration.

Movement

In water, mudskippers swim like fish, moving their body in a sinuous pattern. The closely placed stalked eyes and the snout are, however, above the water because the expanded opercular cavities are kept filled with air.

A mudskipper (*Boleophthalmus boddaerti*) leaves a characteristic trail as it moves through the mud of the intertidal zone.

Mudskippers are also able to float with their body upright by slowly and rhythmically flexing their tail sideways and using their pectoral fins as stabilizers. Thus, when they are swimming mudskippers appear to be hovering over the water surface.

On land, mudskippers do not alternate their limbs in a time sequence as do the lower tetrapods (vertebrates with four limbs), but simultaneously move both pectoral fins while keeping their body rigid. Thus, they hop or jump rather than walk. Movement on land leaves a characteristic trail of a dragged body with marks on either side where the pectoral fins are inserted into the ground. They can also climb rocks or mangrove tree roots. Two structures which enable them to perform these movements are the pectoral and pelvic fins. The pectoral fins, with muscles at the medial end next to the body, act like crutches, while the fused pelvic fins work as a suction device, which is particularly useful for climbing. It is the pectoral fins which enable mudskippers to propel themselves forward and also upward.

Burrows

On the mud flat of the intertidal areas, sometimes on a sloping bank, male mudskippers make burrows or tunnel nests, using their mouth to scoop up mud and then deposit it around the edge of the burrow. As the burrow gets deeper, its walls are made of firmer ground. Thus, the burrow does not easily collapse. Although there is an elevated edge at the entrance, water can still seep into the burrow, which is a near vertical tunnel 1–2 metres deep. A burrow can have more than one side entrance; these tunnels join up with the main one. Continual usage and active burrowing prevent the tunnels from being blocked as a result of tidal flooding.

The water-filled burrows function as nests for spawning, fertilization and brooding. After a successful courtship by a male, a female enters the burrow. Courting behaviour seems to vary among the mudskippers. One type, the leaping behaviour during courtship by males of *Boleophthalmus boddaerti*, is well documented. Young mudskippers, like other fish, hatch out of eggs as tiny larvae. As they grow to adulthood, they take the form of a fish. The females normally take care of the brood.

Male mudskippers, as displayed by *Periophthalmus chrysospilos*, are fiercely territorial, especially after pairing with a female. At low tide, males chase or fight off any possible invaders.

Feeding

The eating habits of the mudskipper species vary. *Boleophthalmus boddaerti* is purely herbivorous, feeding on diatoms, blue-green algae and fungal material which can be found in the upper surface of the mud. *Periophthalmodon schlosseri* is carnivorous, feeding on smaller organisms like crabs and snails and even smaller mudskippers. Both *Periophthalmus chrysospilos* and *Scartelaos viridis* are omnivorous. The diet of *Periophthalmus chrysospilos* is quite similar to that of *Periophthalmodon schlosseri*

DISTINGUISHING FEATURES OF MALAYSIAN MUDSKIPPERS

	BOLEOPHTHALMUS BODDAERTI	PERIOPHTHALMODON SCHLOSSERI	PERIOPHTHALMUS CHRYSOSPILOS	SCARTELAOS VIRIDIS
Maximum length (cm)	13	27	12	15
Difference between sexes	none	none	males smaller, have elongated first dorsal fin with flexible orange spines; in breeding season, undersurface of throat is pale orange	males longer, have white rim around eyes
Colour	6 or 7 oblique blackish bands on lateral sides of body; round spots (pale brown to silvery blue) on body and dorsal fin	dark brown band from eye to posterior region	bluish grey to yellowish brown; round orange spots on whole body; abdomen whitish; broad longitudinal black band on posterior dorsal fin	greenish yellow
Habitat	intertidal zone of mud flats; at high tide enters burrow	intertidal zone of mud flats; at high tide swims at water's edge	mud flats; littoral zone of shores; at high tide climbs up roots of mangrove trees	stays under water near water's edge
Defence	dives and remains submerged	makes 2 or 3 rapid jumps over water surface	makes 2 or 3 rapid jumps over water surface	dives and remains submerged

supplemented with plant material. *Scartelaos viridis* lives on food such as algae and nematodes that can be found within the top layer of the mud.

The latter types are predators, but mudskippers are also prey for snakes and birds, such as kingfishers, which also live in the coastal mangrove areas. Mostly the mudskippers scamper away to avoid being caught by their predators and enemies, but burrows also function as a refuge from predation.

As a food source

In Malaysia, mudskippers are not usually eaten and there is no organized fishing industry connected to mudskippers for local consumption or for export. However, in Taiwan and Japan where mudskippers are considered a delicacy, they are among the highest priced fish.

Mudskipper burrows with raised rims, a result of the digging action used by mudskippers to make their burrows.

A composite of Malaysia's four mudskipper species in their mangrove home

1. *Scartelaos viridis*
2. *Boleophthalmus boddaerti*
3. *Periophthalmodon schlosseri*
4. *Periophthalmus chrysospilos*

Fused pelvic fins enable mudskippers to climb tree roots.

Freshwater crabs, prawns and snails

Malaysia has more than 90 described species of freshwater crabs, mostly endemic species, and also more than 40 species of freshwater prawns, some highly prized in the restaurant trade. Malaysian freshwater snails belong to two groups: the Prosobranchia, with the familiar spiral shells, and the air-breathing Pulmonata, or lung snails. Some are of economic significance as carriers of parasites which can be transmitted to people.

The semi-terrestrial punice crab (*Johora punicea*), discovered only in 1985, is a strikingly coloured species found only on Pulau Tioman. This specimen carries the large eggs characteristic of true freshwater crabs.

The large mountain river crab (*Isolapotamon kinabaluense*) is found only on Mount Kinabalu, Sabah. Members of this genus are usually distinctively patterned and are found in large, fast-flowing, boulder-strewn rivers in Borneo.

This walking stick made by a Mah Meri carver features a beautifully carved prawn on the handle.

The swamp terrestrial crab (*Geosesarma peraccae*) (top) is common in Malaysian lowlands. A spectacular species is *Geosesarma gracillimum* (bottom) from northern Sarawak.

Classification of crabs and prawns

Crabs and prawns are decapods, crustaceans with ten legs (five pairs). They are the most successful of all aquatic arthropods (animals with jointed legs), and some 8,000 species are known throughout the world. Many species have moved into freshwater habitats or even onto dry land. They have become remarkably diverse and successful, adapting so well that their life cycle has become totally independent from the sea where their ancestors evolved.

Freshwater and terrestrial crabs

Several families of crabs rely completely on fresh water though not all actually live in water. Many species have become fully terrestrial, digging burrows and wandering far from permanent water bodies. Some have even become cave dwellers, spending all their life in darkness. Three families—Potamidae, Gecarcinucidae and Parathelphusidae—no longer need to return to the sea to breed. Their large eggs hatch directly into juvenile crabs, eliminating a marine larval phase. However, female crabs take care of their young for many days before releasing them. Many members of another family, Grapsidae, live in fresh water, but some still return to the sea to breed. One genus, *Geosesarma*, has become the newest 'full-time' freshwater occupant, completing its life cycle on land or in fresh water.

More than 90 freshwater crab species have been described in Malaysia and there are undoubtedly many new species awaiting discovery in the unexplored forests, especially in Sabah and Sarawak. Even in the relatively well explored Taman Negara, two new species were discovered in 1992. Crab taxonomy is highly dependent on subtle but important differences in the reproductive organs, male abdomen and mouthpart structures. Thus, identification in the field is not easy.

Several larger freshwater crab species are eaten by local people. However, these crabs should be thoroughly cooked as many species are hosts of the disease-carrying lung fluke *Paragonimus*.

Freshwater prawns

There are two main families of freshwater prawns in Malaysia, Palaemonidae and Atyidae. The clawed prawns of the genus *Macrobrachium*, the dominant genus of palaemonids, are easily recognized by the well-developed pincers of the males; in some species the pincers are longer than the body. Most torrent prawns of the family Atyidae belong to the genus *Caridina*. These are usually only a few centimetres long, but are often present in large numbers. Their first two pairs of pincers are very small, delicate and fringed with long hairs used for filter feeding.

Many freshwater prawn species have a planktonic phase in their life cycle. Some species (including the commercially important *udang galah* (*Macrobrachium rosenbergii*)), have larvae which must develop in the sea; the young then migrate inland after their larval development is complete. Others, such as the very plentiful paddy prawn (*M. lanchesteri*), have planktonic larvae which can develop in fresh water. A few highly adapted species, however, have large eggs which hatch into very advanced larvae without a planktonic phase.

Malaysia has about 25 species of palaemonid prawns and about 15 species of Atyidae. Of interest are two new species of freshwater Alpheidae (snapping prawns), an otherwise marine or estuarine family. Clearly, many new species are still awaiting discovery in Malaysia's rainforests.

All the larger *Macrobrachium* species are eaten by man, although most forest species only by forest people. However, the giant freshwater prawn (*M. rosenbergii*) is a commercially important species which is extensively cultivated in Southeast Asia.

The giant torrent prawn (*Atyopsis moluccensis*) lives in fast-flowing rivers with a rocky substrate. It is much sought after for the aquarium trade.

This freshwater snapping prawn (*Alpheus cyanoteles*) is known only from southern Johor. It lives in fast-flowing streams of freshwater swamp forests.

FRESHWATER SNAILS OF MEDICAL IMPORTANCE

	SPECIES	HABITAT	FOOD	SHELL	SIZE (mm)
Subclass Prosobranchia					
Family Ampullariidae	*Pila ampullacea*	ponds, paddy fields, sluggish streams	aquatic weeds, algae	ovate-conical	55–115 × 55–105
	Pila pesmei			ovate-conical	25–26 × 20–35
	Pila scutata		vascular plants, grasses	ovate-conical	25–50 × 20–30
Family Pomatiopsidae	*Robertsiella kaporensis*	shallow freshwater streams with submerged vegetation and rocks	algae	ovate-conical	<4
	Robertsiella gismanni				
Family Thiaridae	*Brotia costula*	fresh and brackish water	algae	elongate turreted	45–75 × 19–32
	Melanoides tuberculata			elongate turreted	20–40 × 5–15
	Thiara granifera			ovate-conoidal	12–44
	Thiara scabra			ovoid-conical	18 × 8
Family Viviparidae	*Taia polyzonata*	open ponds with little vegetation	algae and diatoms	ovate-conical	20–35 × 15–25
	Filopaludina sumatrensis polygramma			ovate-conical	20–35 × 15–25
Subclass Pulmonata					
Family Lymnaeidae	*Lymnaea auricularia rubiginosa*	paddy fields, ponds, streams, lakes	algae	conical	25–35 × 15–20
Family Planorbidae	*Indoplanorbis exustus*	paddy fields, ponds, lakes	aquatic vegetation	discoid	20–30 × 10–15
	Gyraulus convexiusculus		algae, small invertebrates	planispiral	5–9 × 1–3

The freshwater snail species of medical importance found in Malaysia include *Pila* sp. (1), *Indoplanorbis* sp. (2) and *Lymnaea* sp. (3).

Freshwater snails

Snails belong to the Gastropoda, the largest class of the phylum Mollusca, and the one which has undergone the most striking adaptive radiation (that is, the evolution of several divergent forms from a primitive and unspecialized ancestor). They have invaded fresh water, and several groups have conquered land with the elimination of the gills and conversion of the mantle cavity into a lung. The Prosobranchia, with the typical spiral shell, have the most primitive type of gill structure. The other two subclasses, the Opisthobranchia and Pulmonata, are probably both derived from the prosobranchs. While opisthobranchs largely live in marine habitat, pulmonates are highly successful land snails and there are also many freshwater forms. The lower pulmonates include both marine and freshwater forms; the freshwater forms come to the surface to obtain air. The higher pulmonates are terrestrial forms in some of which, such as slugs, shell reduction or loss has occurred.

Many freshwater snail species found in Malaysia are of public health importance as they are hosts of parasites causing diseases in people, including paddy farmers working in flooded fields. Among these parasites are liver flukes which can be transmitted to people eating snails not thoroughly cooked.

Members of the Thiaridae family transmit flukes. Two recently discovered snail species of the family Pomatiopsidae, *Robertsiella kaporensis* and *R. gismanni*, though only the size of a grain of rice, are both of parasitological importance as they harbour the Malaysian schistosome. *R. kaporensis* harbours *Schistosoma malaysiensis*, which affects the liver.

Snails of two families harbour parasites which cause dermatitis in paddy farmers. *Lymnaea* snails harbour *Trichobilharzia brevis*, while *Indoplanorbis exustus* harbours *Schistosoma spindale*.

However, some snail species also perform useful functions. For example, some apple snails (*Pila* spp.) eat egg masses of other snails. Thus, although considered an agricultural pest as they eat crops such as paddy, they are also a form of biological control for other snails. While the commonest species in Malaysia, *Pila scutata*, *P. ampullacea* and *P. pesmei*, are of health importance as they carry parasites, these snails are eaten by Malaysians and some people believe they have medicinal value.

Pomacea canaliculata

Lymnaea auricularia rubiginosa

Indoplanorbis exustus

Robertsiella sp.

Jahut carvings

Hantu ketam, the crab spirit, which lives in rivers and chases and bites anyone who dives near him. Its bite causes swelling in the victim.

Hantu udang, protector of prawns, whose permission must be sought before fishing. Defiant fishermen will find their stomach swollen with worms.

Hantu siput, protector of snails. Permission is needed for catching snails; otherwise, the eater will have a stomachache.

Snail motifs

Handicraft designs using snail motifs in different mediums: a batik design (top) and a basketware design (bottom).

Fiddler, soldier and hermit crabs

Malaysia's very long coastline has extensive intertidal areas very rich in food resources, with plankton brought in by the incoming tide. There is an abundance of crabs on the boundary where the land meets the sea, including fiddler, ghost, soldier and sand-bubbler crabs belonging to the group Brachyura ('true' crabs) with four pairs of legs, as well as hermit and coconut crabs of the Anomura ('false' crabs) with three pairs of legs.

Intertidal zone crabs

The dominant groups of true crabs living in the intertidal zone belong to the family Ocypodidae. Best known are the fiddler, ghost, soldier and sand-bubbler crabs, effectively terrestrial crabs which breathe atmospheric air. If kept under water, they will drown unless the water is very well oxygenated. During high tide, these crabs reside in burrows, surviving in an air bubble trapped underground. Only when the tide is out do they emerge and forage for food. Most have tufts of fine hairs on their belly (thorax or abdomen) which absorb water from the damp sand by capillary action. This helps to prevent desiccation and keeps their gill chambers filled with water. Able to sense the incoming tide well before the water reaches them, they usually feed until the last moment and only start burrowing just before they are inundated.

Ghost crabs have few predators, especially at night. Fiddler, soldier and sand-bubbler crabs have many enemies, including kingfishers (the most serious), mudskippers and snakes. These intertidal crabs are very important in the beach ecosystem as their constant burrowing ensures buried nutrients are continually exposed and recycled as well as keeps the beach clean and free of disease.

Fiddler crabs

Fiddler crabs (genus *Uca*) are represented by a large number of species in Malaysia, including common sandy shore species such as *Uca vocans* (with a grey carapace and large, granular, orange claw) and *U. annulipes* (black carapace with white lines and a large, smooth, whitish orange claw). The most spectacular species is probably the aptly named bright red *U. rosea* found in the mangroves. Unlike females, which have small claws, male fiddler crabs have one greatly enlarged pincer which is used for fighting (for territory as well as females) and in courtship rituals. Their small pincer is used for feeding. In female crabs, both pincers are small.

Ghost crabs

Ghost crabs (genus *Ocypode*) live just above the supralittoral zone, the part of the shore which is inundated only by the highest spring tides. Their common name comes from their ghostly, greyish white carapace and generally nocturnal habits. The most common species is the horn-eyed ghost crab (*Ocypode ceratophthalma*), which has distinctive horn-shaped stalks above its eyes. By day, ghost crabs dig deep burrows in the sand to escape the hot sun and diurnal predators. At night, they forage at the edge of the water for any carcasses washed ashore. They have also been known to hunt small fish and crabs at the water's edge, pulling them onto dry land with their powerful pincers. In areas where sea turtles breed, ghost crabs can take a terrible toll of the young turtles as they emerge from their nests and dash for the sea. Ghost crabs are very elusive as they are among the fastest of all crabs, especially when running sideways, and have excellent eyesight.

After sifting the sand around their burrows for food, the sand-bubbler crabs leave behind a distinctive pattern of sand pellets.

Soldier crabs

Soldier crabs (genus *Dotilla*) are not only smaller than ghost crabs, but are also more peaceful animals, feeding on the organic matter and plankton deposited on the sand after each high tide. They emerge en masse during low tides and forage for food near the water's edge (*Dotilla myctiroides*) or on the dry sand (*D. wichmanni*). It is this habit of moving about in large numbers which gives the crabs their common name of soldier crabs. *D. myctiroides* feeds by scooping sand into its complex mouthparts, which then sort out the food

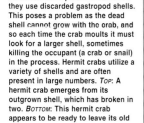

Hermit crabs are unusual in that their abdomens are soft and unarmoured. To protect themselves they use discarded gastropod shells. This poses a problem as the dead shell cannot grow with the crab, and so each time the crab moults it must look for a larger shell, sometimes killing the occupant (a crab or snail) in the process. Hermit crabs utilize a variety of shells and are often present in large numbers. *Top*: A hermit crab emerges from its outgrown shell, which has broken in two. *Bottom*: This hermit crab appears to be ready to leave its old shell to hunt for a new one.

The courtship ritual of the fiddler crab

The male fiddler crab has one enlarged pincer which is used only for signalling—for territorial or mating purposes. This signalling procedure is very complex. Shown here is the vertical display, the initial stage of the mating ritual, before an individual female is identified. Starting with the claw in front of the face, the crab then swings it back to an open position before rotating it up and down.

Colourful crabs of the Malaysian shores

The common ghost crab (*Ocypode ceratophthalma*) is easily recognized by its stalked eyes, which give it an especially eerie appearance. This scavenger is among the fastest of all crabs.

A male common fiddler crab (*Uca vocans*), with a grey carapace and large, granular, orange claw, which is found on many sandy, exposed shores. Females have two small pincers.

The common soldier crab (*Dotilla myctiroides*) has characteristic green eyes and very elongated pincers. It appears in large numbers at the water's edge at low tide.

The common sand-bubbler crab (*Scopimera tuberculosa*) gets its name from its habit of depositing sand pellets which have been sifted for food near its burrow.

The common shore land hermit crab (*Coenobita rugosus*) is a scavenger. In areas where there are no human activities, it moves about during the day.

The coconut crab (*Birgus latro*), the largest land arthropod in the world, is widely distributed in the Indo-West Pacific, preferring small, isolated islands where there are few predators.

from the sand. The inorganic matter is rolled into small pellets which the crab neatly removes with its delicate pincers (which have spoon-tipped fingers) and deposits on to the sand. Each species leaves behind its own distinctive pattern of pellets.

The most remarkable adaptation of *Dotilla* crabs is their ability to breathe with their legs. The shell of the largest leg segment is especially thin and is full of blood vessels. This special structural adaptation supplements the gills and helps to increase oxygen intake, a useful feature for an active animal.

Sand-bubbler crabs

Sand-bubbler crabs (genus *Scopimera*) resemble soldier crabs, but are smaller and their tufts of water-absorbing hairs are on the thorax, not the abdomen; they also have stiff, black bristles on their legs. They forage close to their burrows, leaving radiating from the burrow opening a very distinctive, bubble-like pattern of sand pellets; it is this pattern of sand pellets which gives them their common name.

Hermit crabs

Southeast Asia has several families of hermit crabs, which live in gastropod shells as they have no shell of their own. Despite their often heavy shell, hermit crabs can drag themselves about with some speed. They are scavengers with a very good sense of smell, quickly attracted to decaying animal or plant matter. Their powerful pincers are able to tear apart most organic matter and so their quarry is easily disposed of. When threatened, they retreat into their thick shells, blocking the entrance tightly with their large pincers and so are almost invulnerable to predators.

Of the intertidal hermit crabs, the genera *Clibanarius* (large, bright orange or blue species with equal-sized pincers) and *Diogenes* (small, dull white or grey species with very unequal pincers) are the most frequently seen. Land hermit crabs (genus *Coenobita*) are common in most supralittoral habitats. They are able to wander deep inland as they have specially adapted eyes and respiratory structures which allow them to survive for long periods out of water. They are completely terrestrial crabs, returning to the sea only to breed.

The coconut crab (*Birgus latro*) breathes atmospheric air, and its gill chamber has been modified into a special lung-like structure. The soft abdomen is rolled under its body and has numerous knobs for protection as the coconut crab does not have a shell.

The endangered coconut crab

One hermit crab species is very different from the others. The coconut or robber crab (*Birgus latro*) is the largest living terrestrial arthropod, found only on small islands where there are few predators: young crabs are very vulnerable although adults are well armed. The young coconut crab lives the same way as the common hermit crab, but when it becomes older it sheds its gastropod shell, tucks its soft abdomen underneath, grows up to 50 centimetres long and 10 kilograms in weight, and develops huge, very powerful pincers. Like the common hermit crab, the coconut crab is a scavenger with a strong sense of smell.

The name 'coconut crab' comes from its love of feeding on coconut flesh, even climbing trees in search of the nuts. The smallest crack is enough to allow the crab's powerful pincers to tear a coconut apart. Its alternative name, 'robber crab', comes from its habit of stealing shiny objects.

Hunted relentlessly as a prized seafood, the coconut crab has been exterminated from many islands, and is an endangered species in most other areas.

A chelicerate, like all spiders, the golden web spider (*Nephila maculata*), the commonest of Malaysia's large spiders, spins an enormous, intricate web among trees and bushes.

CHELICERATES, MYRIAPODS AND ANNELIDS

Giant millipedes found in the Malaysian rainforest can be as long as 25 centimetres.

Chelicerates, myriapods and annelids are invertebrate animals (animals without backbones). Chelicerates and myriapods are arthropods (phylum Arthropoda) (animals with jointed limbs), while annelids are segmented or true worms of the phylum Annelida.

Chelicerates belong to the subphylum Chelicerata. Their body is made up of two major regions: the cephalothorax (fusion of head and thorax) and the abdomen. They are characterized by the structures called chelicerae, a pair of pincer-like appendages situated in front of the mouth.

Among the chelicerates in Malaysia are horseshoe crabs (or king crabs, class Merostomata) and arachnids (class Arachnida), which include spiders (order Araneae), scorpions (order Scorpionida), and ticks and mites (order Acarina). Malaysia has three of the world's four horseshoe crab species; these are animals which have not changed in form for millions of years. Spiders are by far the most diverse and most widely distributed of the chelicerates. Although spiders are usually associated with webs, many Malaysian species do not make webs. These include the most primitive of the country's spiders, *Liphistius* spp., which live in burrows with a hinged door. Some live in mossy banks in forests; others live in limestone caves.

Myriapods are divided into two classes: centipedes (class Chilopoda) and millipedes (class Diplopoda). Centipedes are active predators, hunting small prey such as insects, spiders and worms. They paralyse their prey with venom from their 'poison claws', the modified first pair of legs. In contrast, millipedes are scavengers, feeding mostly on vegetable matter. However, many millipedes have a row of poison glands along each side of the body. Pill millipedes, shorter and stouter than most millipede species, defend themselves by rolling up into a spiral.

Annelids include earthworms (class Oligochaeta) and leeches (class Hirudinea). Both earthworms and leeches are hermaphroditic, each individual having both male and female reproductive organs. Visitors to Malaysia's rainforest often find themselves bitten by leeches, which thrive in the moist, tropical climate.

Despite their diverse backgrounds, chelicerates, myriapods and annelids all contain some members which can be harmful or hazardous to humans. Scorpions, some larger spiders and centipedes are potentially venomous; and, given the opportunity, leeches feed on human blood. But some kinds are beneficial to humankind. Earthworms, through their burrowing habit, help in improving the fertility of the soil. *Tubifex* worms are important as fish food in the aquarium fish industry. Many spiders are useful as they prey on insect pests. The blue blood of the horseshoe crabs is being utilized in medical research.

The bright yellow horned spider (*Gasteracantha arcuata*), one of Malaysia's bizarre spider species.

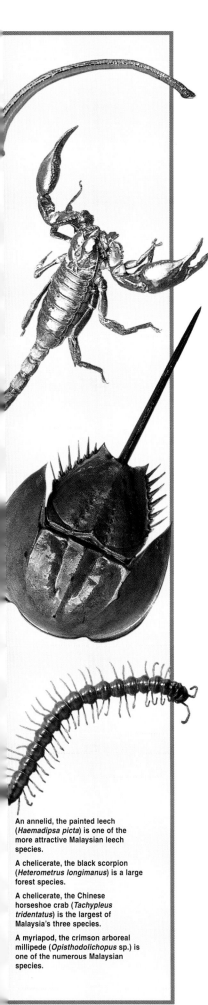

An annelid, the painted leech (*Haemadipsa picta*) is one of the more attractive Malaysian leech species.

A chelicerate, the black scorpion (*Heterometrus longimanus*) is a large forest species.

A chelicerate, the Chinese horseshoe crab (*Tachypleus tridentatus*) is the largest of Malaysia's three species.

A myriapod, the crimson arboreal millipede (*Opisthodolichopus* sp.) is one of the numerous Malaysian species.

Horseshoe crabs

One of the most unusual creatures to be encountered on the shores of Malaysia is a plate-like animal with a long, spine-like tail. Slow moving and totally harmless, it has no offensive tendency, no sharp teeth, no strong pincers and when caught, will merely flick its spiny tail about slowly. This creature is the horseshoe crab, of which Malaysia has three of the world's four living species.

dorsal view

ventral view

A crab that is not a crab. Viewed from above, the reason for calling this creature a horseshoe crab is obvious, but its true relationship, to spiders and scorpions, is revealed when viewed from the underside and its six pairs of legs can be seen.

The book gills of the horseshoe crab help the animal breathe. The structure of the gills and mouthparts tell zoologists that horseshoe crabs are not really crustaceans, but are more closely related to spiders and scorpions.

Classification

Despite being commonly known as 'crabs', horseshoe crabs (*belangkas*) are not crustaceans at all, but belong to the subphylum Chelicerata, which also includes spiders, mites and scorpions. Although both crustaceans and chelicerates are arthropods (invertebrates with jointed legs), there are significant differences between them. Chelicerates do not have mandibles to crush their food, and they also lack antennae. Horseshoe crabs, which are sometimes also called king crabs, are the only truly marine members of the Chelicerata, and have a very long lineage. Ancestors of horseshoe crabs are believed to have evolved 550 million years ago, with the typical horseshoe crab form evolving 150 million years later. For the last 400 million years, horseshoe crabs have hardly changed in structure. For this reason, many zoologists regard them as living fossils.

Although a successful group of animals in the past, modern horseshoe crabs are represented by only four species in three genera. Of these, three species, belonging to two genera, are found in Malaysia. These are the mangrove horseshoe crab (*Carcinoscorpius rotundicauda*) (the smallest living horseshoe crab), the coastal horseshoe crab (*Tachypleus gigas*) and the Chinese horseshoe crab (*T. tridentatus*). The only other living species is the American horseshoe crab (*Limulus polyphemus*), which is found on the east coast of North America.

While the mangrove and coastal horseshoe crabs occur throughout Southeast Asia, the Chinese horseshoe crab occurs mainly in Chinese and Japanese waters. In Southeast Asia, it is known only from northern Sabah. The three species can be distinguished from one another by differences in size, colour, shape of tail and number of spines.

Breeding

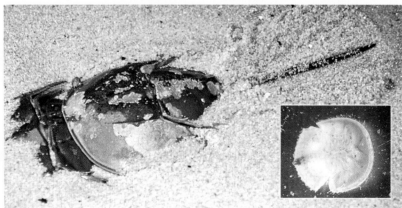

Male horseshoe crabs are dragged ashore by the much larger females and mate in the upper zone of the beach. *INSET*: The young of horseshoe crabs resemble adults except that they have no tail and are bright green.

Horseshoe crabs breed during spring tides, when the effects of the moon and sun exert maximum influence on the seas, causing the highest and lowest tides. It is only during such tides that the supralittoral zone (the upper part of the shore not normally inundated by water) is covered by the sea for a few hours. As the females start to move ashore during high tide, they are intercepted by the males which grip the opisthosoma of the females. The much larger females drag the males ashore, where they dig the soft sand or mud and lay their bead-like eggs, which are immediately fertilized by the males. Covered with sand by the females, the eggs stay buried in the supralittoral zone until the next spring tide when the water level is high enough to inundate them. Horseshoe crabs do not have a free-swimming larval phase; the eggs hatch directly into a miniature version of the adults, except they are bright green and have no tail. Because of their appearance, young horseshoe crabs are often called 'trilobite larvae' although they are not related to the extinct trilobites. At the next spring tide, the young burrow out of the sand and swim into the open sea, where most fall prey to predators.

Structure

The body of a horseshoe crab is divided into two parts: the front half (prosoma) contains its two pairs of eyes, the mouthparts and legs, while the posterior half (opisthosoma) contains the book gills used for respiration as well as the digestive, nervous and reproductive systems.

Horseshoe crabs are generalists, feeding on any form of animal (and sometimes vegetable) matter, using their first pair of legs for feeding. As essentially sublittoral animals, they live close to the seashore, scavenging for food as well as actively forcing open small bivalves.

Contrary to some reports, the spine-like tail is not used to protect the animal against aggressors. Rather, it is used to right the animal when it is inadvertently turned upside down. By digging its long spine into the sand, it obtains enough leverage to twist itself right side up. Only its thick shell protects the horseshoe crab, although adult specimens generally have few enemies, except perhaps sharks or turtles. There have been reports of horseshoe crabs being poisonous, but such incidents were probably a result of the crabs eating something poisonous, not because they produce toxins.

Females are easily distinguished from males, not only by their larger size (males are usually only half the size of females), but also by the frontal margin of the prosoma. In males, it appears sinuous from the dorsal view as the ventral surface is depressed (concave). By contrast, in females it appears gently convex from the dorsal view as the ventral surface is flat. During reproduction, the male clings onto the back (opisthosoma) of the female. The ventral concavity on its frontal margin helps the male attach itself to the convex surface of the female's back.

Eggs as a food resource

Horseshoe crabs have always been a valuable resource for the people living along the shores of Malaysia. Their eggs are regarded as a delicacy, and it is still common to see horseshoe crabs being sold live in local markets. It is important to remove the eggs very carefully from the ovaries and to avoid puncturing the gut as fluid in the gut may contain poisonous substances and hence contaminate the eggs. Traces of tetrodotoxin (a powerful nerve toxin) have sometimes been found in horseshoe crab eggs.

Medical research

In recent years, horseshoe crabs have become very valuable for medical research because their blue blood is one of the most potent detectors of gram-negative bacteria (which cause the majority of the dangerous bacterial diseases in man) and their lethal by-products. When this blood comes into contact with the endotoxins produced by such bacteria, even in very minute quantities, it immediately clots. Test kits made from the blood of horseshoe crabs have become extremely important in detecting and analysing diseases as well as testing if laboratory ware is really free of bacterial contamination.

Unfortunately, the increasing medical importance of horseshoe crabs has serious drawbacks. To prepare the bacterial test kits, blood must be collected from live animals. If the blood is collected carefully (that

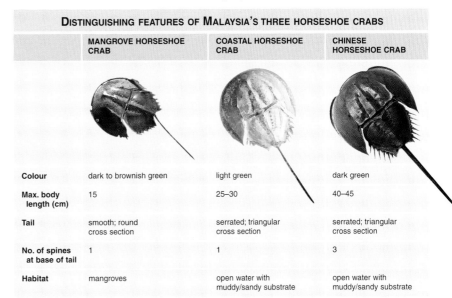

DISTINGUISHING FEATURES OF MALAYSIA'S THREE HORSESHOE CRABS

	MANGROVE HORSESHOE CRAB	COASTAL HORSESHOE CRAB	CHINESE HORSESHOE CRAB
Colour	dark to brownish green	light green	dark green
Max. body length (cm)	15	25–30	40–45
Tail	smooth; round cross section	serrated; triangular cross section	serrated; triangular cross section
No. of spines at base of tail	1	1	3
Habitat	mangroves	open water with muddy/sandy substrate	open water with muddy/sandy substrate

is, no major organs are punctured and the animal is not badly injured during capture) and in not too great a quantity, the animals are able to survive after release. However, unscrupulous people drain almost all the blood from crabs and then release the almost dead animals as a gesture of 'goodwill'. As a result, in some parts of Asia large species, such as the Chinese horseshoe crab, have been overexploited for their blood and have become very rare. Efforts are now being made to culture the crabs as well as to artificially produce the active ingredients of the horseshoe crab's blood by genetic engineering. Hopefully, these steps will help to conserve horseshoe crabs in the wild.

Toughness and longevity

Their bioactive blue blood aside, the toughness of horseshoe crabs is legendary. They are able to survive for days in fresh water or in hypersaline sea water, they can withstand prolonged periods of desiccation, and they can live out of water for several days. Whereas high temperatures kill many marine animals in minutes, horseshoe crabs can survive many hours. They can also be starved for several weeks with no apparent harm and, if necessary, can feed on almost any plant or animal matter.

Yet, horseshoe crabs have never been especially diverse as a group. There were a large number of species 300–360 million years ago, but since then there have been a mere handful of species present at any one time. However, these few species have survived for much longer periods than other species, with each surviving for millions of years seemingly unchanged. If longevity is used as a measure of biological success (along with the more usual factors such as diversity and the number of habitats colonized), horseshoe crabs are surely the Methuselahs of the animal world. As one palaeontologist has so aptly said, horseshoe crabs are a 'timeless design'.

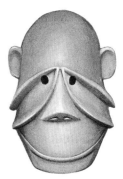

Moyang belangkas, one of the Mah Meri spirit figures which are used by shamans to extract the source of a patient's illness. Having absorbed the disease, the carving is then discarded.

The long tail of the horseshoe crab is not used, as might be imagined, for defence, but rather as a lever to turn itself over should it be turned upside down on land.

An inquisitive group of holiday-makers holding horseshoe crabs, some of a large number seen on the beach at Lumut, Perak. These horseshoe crabs were later returned to the sand.

Spiders

In existence since the Devonian period 400 million years ago, there are today probably more than 40,000 known species of spiders in the world. The number of species in Malaysia is not known, but most certainly runs into the hundreds, and many more species are waiting to be discovered and named.

A nursery web spider (*Thalassius albocinctus*), which lives near water. This spider has been reported to attack fish for food.

In their handicrafts, the Iban of Sarawak use many animal motifs, such as this stylized spider motif.

Classification

Spiders are members of the phylum Arthropoda, that is, animals with jointed limbs. They belong to the class Arachnida which is characterized by the possession of four pairs of legs and two major body regions: the cephalothorax (a fusion of head and thorax) and the abdomen. Other members of the class Arachnida are the scorpions (order Scorpionida), daddy-longlegs (order Opiliones), and mites and ticks (order Acari). The spiders belong to the order Araneae. They differ from other arachnids by the possession of a conspicuous 'waist' (a constriction between the cephalothorax and the abdomen), venomous fangs (jaws called chelicerae), and silk-producing glands in the abdomen.

Size

Spiders in Malaysia, as elsewhere, come in a great variety of sizes, forms and colours. The biggest spiders are the Mygalomorphs, such as members of the genus *Lampropelma*, which are commonly referred to as 'bird-eating spiders'. The commonest large spider in Malaysia is the golden web spider (*Nephila maculata*), which spins an enormous web among trees and bushes.

Female spiders are normally significantly larger than the males of their own kind, a condition which is known as sexual dimorphism. Differences between male and female spiders are also reflected in their different body colour and colour pattern. Of the

Ant mimicry

Some spiders look like ants, though the function of this ant mimicry is unknown. The weaver ant-like jumper *Myrmarachne plataleoides* (male, above left; female, above right), which resembles the weaver ant (*kerangga*) (*Oecophylla smaragdina*), is found on trees and shrubs inhabited by weaver ants. Another, and perhaps even more striking, *kerangga* mimic is the ant-like crab spider (*Amyciaea lineatipes*) (below right) whose abdomen resembles the head of the weaver ant. The resemblance is accentuated by the presence of two dark spots on the abdomen which look like the ant's eyes. This crab spider preys on the weaver ants!

large spiders, the females of *Nephila maculata* measure 30–50 millimetres long, with the legs spanning 150 millimetres or more; the males are very much smaller, only 5–6 millimetres long. Likewise, of the small spiders, females of *Misumenops nepenthicola* are 6 millimetres long while the males attain only half that size.

Habitat

Most spiders inhabit terrestrial habitats—forests, plantations, gardens, caves, bushes, and even nests of ants and termites—as well as the interior of buildings. A few are, however, aquatic or semi-aquatic, inhabiting ponds, streams and other water bodies (for example, some lycosids and pisaurids).

Bizarre species

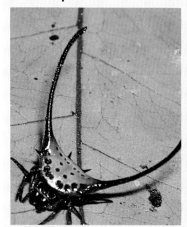

Among the bizarre-looking spiders are members of the genus *Gasteracantha* (such as *G. arcuata,* above left), commonly known as spiny-bodied spiders because their bodies are beset with spines. The most bizarre of them are, perhaps, those spiders which look like ants or resemble bird-dung and the like. As characteristic, but in a different way, are the signature spiders (species of the genus *Argiope*, such as *A. versicolor* above right) which construct zigzag bands of silk into the middle of their webs.

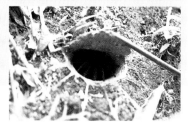

Biologically, the most interesting of the Malaysian spiders are members of the genus *Liphistius*, such as *L. desultor*. These are by far the most primitive spiders and are often called 'living fossil spiders' as they are related to fossil forms dating back more than 200 million years to the Carboniferous period. There are two groups of *Liphistius* spiders in Malaysia, one inhabiting the mossy banks in forests, and the other confined to limestone caves. These spiders live in burrows with a hinged door. Radiating from the mouth of the burrow are several lines of silk thread which serve as communication lines to notify the spider of the imminent arrival of prey.

The *Nepenthes* crab spider

The *Nepenthes* crab spider (*Misumenops nepenthicola*) lives in pitchers of *Nepenthes* plants, staying at the upper zone of the pitcher where it preys on insects attracted there by the sugary fluid. In the face of danger, its special body covering enables it to retreat into the digestive fluid of the pitcher.

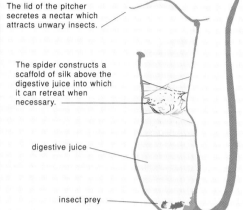

The lid of the pitcher secretes a nectar which attracts unwary insects.

The spider constructs a scaffold of silk above the digestive juice into which it can retreat when necessary.

digestive juice

insect prey

A pitcher plant

There is at least one species, the *Nepenthes* crab spider (*Misumenops nepenthicola*), which stays exclusively within the pitcher of *Nepenthes* plants. There are perhaps other crab spiders that inhabit the plant pitchers, such as *Thomisus nepenthiphilus*, reported in Sumatra. Wherever they live, spiders generally cannot withstand dry air; they perish through losing water by evaporation.

Diet

All spiders are predators, feeding mostly on insects. This ubiquitous habit makes them friends of the farmers. Indeed, they can be an effective biological control agent against agricultural insect pests. As most spiders are typical ambush (sit-and-wait) predators, in times of scarcity their food supply may not be constant. However, they can survive without food for considerable periods.

Enemies

Despite their role as a predator in the food web, spiders too have their fair share of enemies. They fall prey to small mammals, birds, reptiles, amphibians, beetles, ants, centipedes and even other spiders. Some wasps lay eggs in living crab spiders. For survival, some spiders have evolved ingenious mechanisms to evade predators. For example, the Asiatic spider (*Cyclosa mulmeinensis*) builds a number of pseudo-platforms on its web with remains of its prey. These platforms give the illusion of being occupied by a spider, and hence fool would-be spider-hunting predators.

Although spiders are generally harmless and even beneficial to humans, some people do suffer from arachnophobia—an unreasonable fear of spiders.

The fighting spider

Spiders have not been normally kept as pets. In Malaysia, at least in the not too distant past, males of the fighting spider (*Thiania bhamoensis*) were

Thiania bhamoensis, the fighting spider popularly kept by children.

collected and kept by children in matchboxes or the like for 'fighting' bouts. These male spiders are acclaimed for their sparring behaviour, more so perhaps than the Siamese fighting fish.

Making a web

The spider begins its web by making a scaffolding. It then adds spokes from side to side. These spokes are of dry silk, but those added next, around the centre of the web, are very sticky, to catch prey.

Spider silk

Spider silk is elastic and resilient. Depending on the species, it ranges in thickness from 0.0005 millimetre to 0.005 millimetre. Although silk production is normally associated with the silkworm *Bombyx mori* (a moth), spider silk also has commercial applications. It is used for making cross hairs in optical instruments, such as microscopes, telescopic gun sights and surveying equipment.

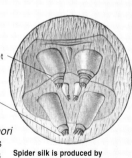

spinneret

spigot

Spider silk is produced by glands inside the spider's abdomen connected to spinnerets on the underside of the spider. Near the rear end are the glands which produce spider silk. The silk is extruded through tiny nipples called spigots.

It is believed that spider silk fibres are also ideal material for incorporation in the making of bulletproof vests, lightweight helmets, parachute cords and the like, particularly for military application. Historically, spider silk was used in making fishing nets, bird snares and small pouches. Cobwebs have found use as bandage material and even as artists' canvas!

1. *Leucauge* sp., a big-jawed spider which lives in low vegetation. 2. The lawn wolf spider (*Hippasa holmcrae*), which makes its web on the ground. 3. The golden web spider (*Nephila maculata*), the most common of the large spiders in Malaysia which spins an enormous web.

Scorpions, centipedes and millipedes

Although scorpions, centipedes and millipedes are not related in a scientific sense, people tend to link them together as in rural areas of Malaysia they are all found around habitated areas and all have the reputation of being dangerous. Although scorpions have poisonous stings and centipedes have poisonous bites, millipedes are not poisonous. Another incorrect perception is that centipedes and millipedes differ only in the number of legs; in fact, there are several differences between them.

The young of the scorpion are born alive and carried about by the mother on her back and limbs.

The scorpion's sting

The scorpion's sting is operated by opposing muscles at the base of the sting which contract and relax, forcing the sharp tip into the victim's tissue. The poison stored in the poison gland is forced down and out of the tip of the sting by muscles around the gland.

muscle contracts
muscle relaxes
poison gland

Scorpions

Scorpions (*kala jengking*) form the order Scorpionida of the class Arachnida. They have a flat, narrow body with two lobster-like claws, eight legs and a segmented tail. Scorpions are generally secretive and nocturnal, hiding by day under logs and stones and in burrows in the ground. Young scorpions are born alive, and in their early stages the mother carries them about clinging to her back and limbs. The young remain on the mother through the first moult, which occurs after about one week. The young scorpions then gradually leave the mother and become independent. It takes about one year for them to reach the adult stage.

All scorpions carry a sting at the end of the tail containing a neurotoxin which affects the nervous system of their victim. Although the sting is severe, none of the five Malaysian species is known to be dangerous to man. To sting, the scorpion raises the post-abdomen over the body so it is curved forward, and uses a stabbing motion to inject poison into its victim.

Scorpions feed on insects and millipedes, partially digesting their food outside the body before ingestion.

LEFT: Scorpion Man, long famous as the Snake King for his daring adventures with snakes, spent three weeks in 1997 living with 5,000 scorpions in a glass cage in a shopping centre in Kuala Lumpur where he drew many curious visitors.

RIGHT: Among the Iban of Sarawak there is a long tradition of body tattooing, with many designs taken from nature. The tattoo on the throat is known as a scorpion.

Usually a scorpion gives warning of its intent to attack by making a sound. This could be the rasping of the mouthparts against the roof of the mouth, the legs rubbing together or the tip of the tail rubbing on the abdomen.

A legend says that if a scorpion is placed inside a ring of fire, it will commit suicide by stinging itself. The truth is more likely to be that the heat from the fire causes the tail to bend over, giving the impression of a self-inflicted sting.

In Malaysia, the little spotted house scorpion (*Isometrus maculatus*) is the species most frequently encountered, nearly always in dwellings. It is slender and yellowish with a pattern of dark spots. The wood scorpion (*Hormurus australasiae*) is abundant, and is found in the jungle under loose bark or dead and fallen trees. It is black and much flattened, which enables it to creep into narrow crevices.

The black scorpion (*Heterometrus longimanus*) is one of the largest scorpions found in the Southeast Asian region. In spite of its size (7–15 centimetres), it is not considered to be dangerous to humans, and there has been no documentation of the scorpion causing any mortality. This species shuns the light of day, and so it is usually encountered at night.

There are two other species of scorpions in Malaysia, the very slender reddish brown *Lychas scutatus*, which is not rare, and *Chaerilus* sp., which is rather uncommon.

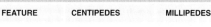

Centipedes

Centipedes (*lipan*), which form the class Chilopoda, are distributed throughout the world, and live in soil and humus, beneath stones, bark and logs. Both centipedes and millipedes have a body composed of a head and an elongated trunk with many leg-bearing segments. Centipedes are fast-running carnivores which can inflict a poisonous bite. Each segment of the body carries one pair of legs, with their bases far apart. Covering the mouth appendages is a large pair of poison claws known as maxillipeds. In centipedes, the last pair of legs is the longest, and in some species they can be used for defence by pinching. Centipedes feed on insects, earthworms and snails, which are located by contact with the antennae or legs and stunned with the maxillipeds before being eaten.

The most conspicuous centipede in Malaysia is the *Scolopendra*, which grows to about 25 centimetres in length and is known to inflict a severe, but not lethal, poisonous bite. It is reddish brown, with 21 pairs of walking legs, the last pair the longest. Another genus of centipede, the *Geophilus*, encountered in houses, is a slender animal 5–7.5 centimetres long, which emits a luminous liquid when touched. This animal need not be feared as its poison claws are too small to penetrate the human skin.

Centipedes lay their eggs in an underground chamber, usually beneath a stone. They then coil around the eggs, guarding them very tenaciously against possible enemies, which include fungi, mites and insects, until they can look after themselves. The larvae of some centipedes have a full set of legs when they hatch; others have only seven pairs of legs, the others being added at the time of moulting.

DISTINGUISHING FEATURES OF CENTIPEDES AND MILLIPEDES		
FEATURE	**CENTIPEDES**	**MILLIPEDES**
Class	Chilopoda	Diplopoda
Antennae	long	short
Body	flat	cylindrical
Defence	vigorous fight	roll into ball
Diet	predatory	herbivorous
Exoskeleton	soft	hard
Legs	long, 1 pair in each segment, bases far apart	short, 2 pairs in each segment, bases close together
Movement	active wriggle; legs on two sides move alternately	slow-moving glide; legs on both sides move together
Poison	poison fangs	sting glands

A centipede (left) has long legs with bases far apart; each body segment has one pair of legs. A millipede (right) has shorter legs, close together at the base; each body segment has two pairs of legs.

Millipedes

Millipedes (*sepah bulan*) are elongated and cylindrical with very short legs, and are commonly known as the 'thousand-leggers'. Each body segment has two pairs of legs. They are secretive and largely shun light, living beneath leaves, stones, bark and logs, and in soil. Some inhabit old burrows of other animals, such as earthworms, and a few are commensal in ant nests. They live throughout the world, especially in the tropics. In general, millipedes are not very agile animals, most species crawling slowly over the ground. All the legs are involved in producing wave-like movements which help the animal to push through humus, leaves and loose soil.

Many millipedes lay their eggs in a nest they have made from soil and their own droppings. Some stay with the eggs until they hatch; others do not.

The most common millipede in Malaysian gardens is the bright red *Trigoniulus lumbricinus* which measures 2.5–5 centimetres in length. Larger millipedes are found in rainforests, and are either black or brown. The flat millipedes are about 10 centimetres long, and have a pair of projecting plates on each segment, while the pill millipedes (order Oniscomorpha) are short, and look like beetles. They roll up into a ball when disturbed.

Millipedes are primarily herbivorous, feeding mostly on decomposing vegetation. Food is usually moistened by secretions and chewed or scraped by the mandibles. The millipedes usually eat the discarded exoskeleton to aid in calcium replacement. Most millipedes live for several years.

To compensate for the lack of speed in fleeing from predators, a number of protective mechanisms have evolved in millipedes. They protect themselves by means of sting glands along the sides of the body. These secrete hydrocyanic acid or benzoquinone. In some species, the last tergite (covering on the dorsal part of the body) is expanded laterally and covers the head when the animal is rolled up.

A centipede seen in Ulu Langat, Selangor.

The pill millipedes (Oniscomorpha) roll themselves into a ball when disturbed.

A millipede photographed at Genting Highlands, Pahang.

Cave centipedes

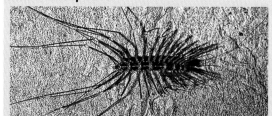

The long-legged centipede *Scutigera* has a short, stout body and 15 pairs of elongated legs. The antennae and hindmost pair of legs are filamentous and are much longer than the body. Often not recognized as a centipede, *Scutigera* is found in caves and under logs, and even sometimes in houses. The cave species are larger, and run very swiftly. They feed on the bat guano on the cave floor.

Leeches and earthworms

Leeches and earthworms are both members of the phylum Annelida. In Malaysia, leeches (class Oligochaeta) are mostly considered a nuisance to both humans and animals, although they have medical uses. Earthworms (class Hirudinea) are appreciated for their beneficial effect on soil fertility.

Rather different from the usual dull brown leeches is this more colourful species (*Haemadipsa picta*) from Sabah.

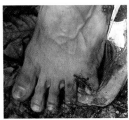

Visitors to the forest often find themselves feasted upon by leeches. Their presence often goes undetected until later when bleeding alerts the hapless victim or, as perhaps in this case, when clothing is removed.

Annelids

Leeches (*pacat*) and earthworms (*cacing tanah*) are both elongate, wormlike animals whose bodies are divided into equal rings or segments and which breathe through the skin or with gills. The name of the phylum is derived from the Latin word *annelus*, 'little ring'. Earthworms, like most annelids, have bunches of horny bristles (chaetae) sprouting from the skin. Leeches, however, do not.

Leeches

Leeches are relatively common animals found in all regions of the world except those covered by snow for most of the year. Nearly three-quarters of the world's leeches are found in freshwater or terrestrial habitats, and vary from a few millimetres to several centimetres long.

Leeches are soft-bodied annelids of the class Hirudinea, always with 34 segments. They are well adapted to a carnivorous diet in place of the detritus-feeding habits of their ancestors. Leeches may be regarded either as external parasites or as rather specialized predators, as they attach themselves to their host only long enough to have one meal at the host's expense. A leech swells by as much as three times its body weight as it takes blood. After the meal, it detaches itself and soon seeks a mate. Leeches can go for months or years between meals. They are hermaphroditic (one animal contains both male and female sex organs), and lay eggs in water. The parents carry the young, from eggs, in a pouch formed by a fold on the lower surface of the body. Sometimes, the young are attached to projections on the underside of the parent. Leeches are known to show advanced parental care.

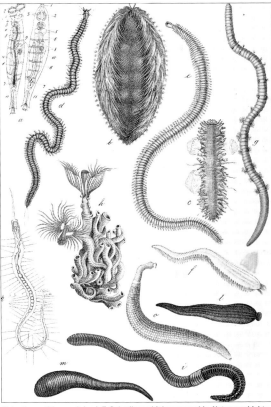

This chromolithograph by J. F. Schreiber, which appeared in *Naturgeschichte* by Dr G. H. von Schubert, published in Germany in the late 19th century, includes drawings of an earthworm (i) and leeches (l, m).

Leeches are mainly found in the tropics and subtropics where, because of their avid blood feeding habits, many species are a scourge to man and animals. An anaesthetic and an anticoagulant (hirudin) are present in the salivary secretions of the leech; these prevent pain and the clotting of blood during the feeding process. Therefore, the wound made by the leech while biting continues to bleed long after it has detached itself from the victim.

In Malaysia, leeches are very common and are subdivided into two major groups: Rhyncobdellae and Arhyncobdellae. The former possesses a piercing organ (proboscis) associated with the mouth; the latter group has jaws armed with teeth. Two groups of leeches belong to the subclass Rhyncobdellae: the Ichthyobdellidae and the Glossiphonidae. The former are found on fishes, turtles and tortoises. Turtles and tortoises are infected with *Ozobranchus branchiatus*, leeches with special adaptations, such as lateral gills, for respiration.

The Glossiphonidae are found on fishes, frogs, tortoises and molluscs. A Malaysian species, *Batracobdella reticulata*, measuring about 5 millimetres by 3 millimetres, is found on vegetation and freshwater mussels. They do not cause much mechanical injury to the host. Many of these species transmit blood parasites found in wild animals in Malaysia. Most do not parasitize invertebrates, but kill them outright, and therefore they can be enlisted as a biological control for snails.

In the past, the medicinal leech (*Hirudo medicinalis*) was used extensively in Europe for blood-letting as this was thought to be helpful in treating almost every illness. In Malaysia, a similar

The leech as a motif

The leech is often used as a motif in handicraft design by the peoples of Sarawak. When used on *parang* (jungle knife) handles and scabbards, both large and small leeches feature in the motif, with a base pattern to bind the leeches together and to provide 'homes' for them. Here are examples of the use of leech designs on a parang handle and scabbard (left), along the border of a traditional Iban skirt (right) and on basketware (below).

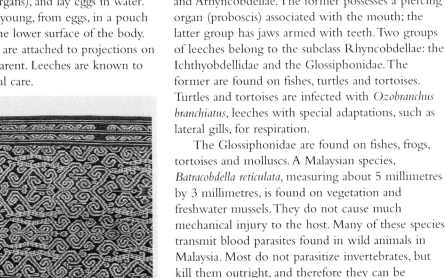

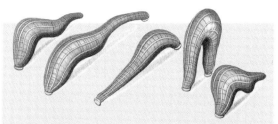

Leech movement

The movement of leeches was vividly described by George Maxwell in his book *In Malay Forests* (1907): *When a leech sees the object of its search there is no further delay—no more bowing and curtseying; it races towards its goal. The head is thrust out as far as it will reach, and the mouth seizes hold of whatever it may touch, a leaf or blade or the bare soil. The body is bent into a great loop that brings the tail up to the head. Then the long body straightens again, and the head is thrust forward once more. Each step is the full length of the body, and the leech covers the ground in graceful sinuosities that remind one of galloping greyhounds.*

species (*Hirudinaria manillensis*), common to swampy places, mainly attacks buffaloes, but has also been recorded in an estuarine crocodile (*Crocodylus porosus*). The true cattle leeches (*Dinobdella ferox* and *Limnatis dissimulata* and *L. maculosa*) have been found in the nose, throat and vagina of horses, cattle and buffaloes. All three species of leeches are a potential danger to man. Tight-fitting clothing acts as a repellent from insect bites, but is not efficient protection against leeches.

The land leeches (*Haemadipsa*) are smaller than buffalo leeches. They live in damp places, hide in crevices during dry weather and emerge to feed on warm-blooded animals when conditions are suitable. Common in Malaysia, they have elastic bodies and the typical looping movements of leeches. They feed on mammals, but have also been recorded on birds and cold-blooded animals. Malaysia's four species of *Haemadipsa*—*H. zeylanica zeylanica, H. subagilis, H. sylvestris interrupta* and *H. picta*—found in moist, cultivated areas, can be an occupational nuisance.

Leeches are also pests to domestic animals as they result in loss of blood, secondary infection from bites, and the transmission of infectious diseases.

Earthworms

About 1,000 species of earthworms inhabit the temperate and tropical regions of the earth. They vary greatly in size; in Malaysia, they are generally about 15 centimetres long. *Pheretima* is the commonest genus of earthworms found in Malaysia. Very large earthworms are known in all continents and several islands, but are unusual and scarce. A giant earthworm (75 centimetres) found on Gunung Lawit, Terengganu, in 1994, which has yet to be identified, is the first record of a large earthworm in Malaysia. The giant earthworms of Ecuador and Australia measure more than 2 metres in length.

The most striking feature of the earthworm is the division of its body into a number of similar ring-like parts called somites or metameres; the linear arrangement of these segments is always limited to the trunk region. Chaetae on the sides of the body, recessed into pits, can be withdrawn or extended at will, like a cat's claws. They are used by the worm to cling to the side of the burrow. Earthworms are hermaphroditic, but cross-fertilization is the rule. Each worm places its ova and spermatozoa in a cocoon which is laid on moist soil. In 2–3 weeks, a single worm, resembling the adult, develops from each cocoon. Some regeneration is possible; the head can generate a new tail, but not vice versa.

Earthworms are well adapted to living in burrows and can force their way through soft earth, but eat their way through harder soil, passing the dirt through the alimentary canal. They prefer loam or partly sandy soil rich in humus. The importance of the working of the soil by earthworms is well known—worms in one acre of land may bring to the surface 18 tonnes of soil per year. Charles Darwin calculated that one acre of land could contain 50,000 earthworms. Earthworms can be used to produce compost, known as vermicompost, from agricultural waste for farming.

As nocturnal animals, earthworms lie in their burrows during the day and come out at night to feed on dead, organic matter, which is digested and deposited as worm castings. They lack respiratory organs, and gas exchange takes place over the entire body surface. If the skin is not kept moist at all times, the earthworm will suffocate.

Occasionally, earthworms are found on paved surfaces after rain because as rain fills the burrows and the earthworms start to suffocate, they leave the burrows for oxygen. Those stuck to a paved surface when the sun emerges soon die as ultraviolet radiation gives them a fatal sunburn in a few minutes. Earthworms are prey to birds, reptiles and some mammals, and are also used as bait by anglers.

The *Tubifex* is a member of a large family of aquatic oligochaetes. It is about 3 centimetres long, red in colour, and coils tightly when disturbed. It lives in the bottom of rivers in thick mud, and emerges with the body undulating gently in the water for respiratory exchange. *Tubifex* worms feed by passing mud through the gut, to digest the adhering organic matter. This is the worm that is sold in shops as food for freshwater aquarium fish.

Unlike insects, earthworms have no legs and so are rather snake-like in appearance.

The early bird catches the worm.

This popular saying probably originates from the earthworm's early morning return to its burrow after its nocturnal activities. Only early risers, such as this Eurasian tree sparrow (*Passer montanus*), have a chance to capture this prey.

How an earthworm moves

Two sets of muscles help earthworms move through the soil. One set is the clearly visible rings which make up the worm's body. The other set is along its length. To move forward, the earthworm elongates the front part of its body by contracting the ring muscles and expanding the long muscles.

Simultaneously, the hind portion becomes thicker by the reverse muscle action. This is followed by thickening of the front portion and elongating the back part, made possible by the sea of fluid between the muscles and the interior organs. In climbing to the surface, the earthworm uses the chaetae to prevent it from slipping backwards. These chaetae are also used by the earthworm to anchor its tail in the burrow when it emerges to feed at night. This facilitates a quick return to the burrow when danger threatens. Worms are favourite prey of many animals, especially ground-dwelling species such as shrews.

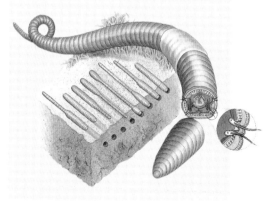

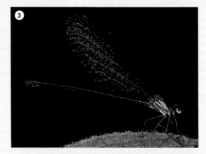

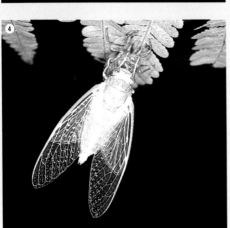

1. Rajah Brooke's birdwing (*Troides* (*Trogonoptera*) *brookiana*), considered by some to be the world's most beautiful butterfly, was first described in the 1850s by the well-known naturalist, Alfred Russel Wallace, who named it after James Brooke, the Rajah of Sarawak from 1841 to 1868.

2. *Necroscia punctata*, one of the many Malaysian stick insects which depend on their remarkable likeness to sticks as a means of defending themselves against predators.

3. The beautiful damselfly, *Vestalis amoena*, one of the more than 200 species of dragonflies and damselflies found in Malaysia. These insects are found near water, as it is there that they lay their eggs and the nymphs live.

4. After a cicada emerges from its long maturation period underground, it climbs up a tree trunk and emerges from its nymphal case. This cicada is drying off after shedding its case, which is beneath the cicada.

5. *Phyllium giganteum*, one of the extraordinary leaf insects which mimic leaves so well that it is very difficult to distinguish them from real leaves. Even the insects themselves are sometimes fooled, and try to eat another insect instead of a leaf.

RIGHT: Cross section of a beehive showing the life cycle of a bee. The egg hatches into a larva, which is fed by a worker bee before its cell is sealed. After the larva has grown it turns into a pupa before emerging from the hive as a fully formed adult.

...kipper butterfly (*Plastingia naga*) and its ...erpillar. One feature which distinguishes ...per butterflies (Hesperiidae) from 'true' ...erflies is their antennae, which are often ...t backwards at the apex.

...common rose (*Pachliopta aristolochiae*), ...riking butterfly whose bright colours ...ably warn potential predators of its ...asteful nature.

INSECTS

Insects are masters in many environments. They inhabit almost every kind of habitat, both terrestrial and aquatic. Their success is due largely to their ability to exploit an enormous variety of food resources: plants and plant juices, small animals, the blood of vertebrates, humus, etc. Both beneficial and harmful insects exist. Being the world's most important herbivores, a large number are crop pests, much dreaded by farmers. Some are vectors of diseases, such as malaria, dengue and filariasis. Nonetheless, insects play a vital role in the breakdown and recycling of dead vegetation and animal matter. They are the sole food of a wide range of vertebrates, and some are providers of silk, honey, wax, pharmaceutical substances, and even food for humans.

Insects are the most successful and probably the commonest of all living things on earth. About a million insect species have been recorded (about three-quarters of all animal species), but the actual number may be 2–4 million species. The number of insect species found in Malaysia is not known.

In just 10 articles in this volume, only a small number of the many insect classes can be described. Among Malaysia's beautiful butterflies (about 1,000 species) and moths (about 10,000 species) are the largest moth in the world and also the butterfly considered by many to be the world's most beautiful. Also very attractive are the dragonflies and damselflies (about 250 species) which are always found not far from water as this is where their eggs are laid. The presence of dragonflies is also an indicator of the quality of the ecosystem.

Malaysia has three types of bees—stingless bees, giant honeybees and hive bees—which all produce honey. However, while stingless bees nest in tree trunks, giant honeybees attach their hives to the branches of very tall trees. Thus, much effort is needed to obtain the honey from these bees. The related hymenopteran species of hornets and wasps are perhaps best known for their fierce stings. However, some wasp species act as biological control agents.

Ants and termites, groups of insects of similar structure but different orders, are found in both forests and homes. Many Malaysian ant species are known for their painful bites, while termites are destroyers of both agricultural crops and buildings.

Much less well known than many of the other insects are the fascinating leaf and stick insects of the rainforest. Their uncanny resemblance to leaves or sticks provides perfect camouflage from predators. Though many species of Malaysia's bugs are very attractive, most are not admired as they are plant pests; some cause enormous economic damage to agricultural crops. Similarly, weevils also destroy crops although some other beetles are beneficial as they devour other insect pests.

Both cicadas and crickets are best known for their songs, though these are produced in different ways. The close relative of the cricket, the grasshopper, is also familiar as a plant eater, but Malaysian species do not cause the serious economic damage known in some other countries.

Although not confined to dwellings, it is with houses that cockroaches, flies and mosquitoes are most often associated as this is where they are seen every day.

Ants are often found resting between the leaf sheath and stem of the climbing bamboo.

Butterflies and moths

*Malaysia hosts a diverse assemblage of butterflies and moths, exhibiting a wide variety of habits and lifestyles. They include what many consider the world's most beautiful butterfly, Rajah Brooke's birdwing (*Troides *(*Trogonoptera*) *brookiana*)*, and the world's largest moth, the atlas moth (*Attacus atlas*).*

Rajah Brooke's birdwing

Although not restricted to Sarawak, the Rajah Brooke's birdwing (*Troides* (*Trogonoptera*) *brookiana*) has become associated with that state as it was first discovered there in 1855 by Alfred Russel Wallace, who named it after James Brooke, Rajah of Sarawak from 1841 until 1868. Wallace described it thus: *It is deep velvety black, with a curved band of spots of a brilliant metallic-green colour extending across the wings from tip to tip, each spot being shaped exactly like a small triangular feather... The only other marks are a broad neck-collar of vivid crimson, and a few delicate white touches on the outer margins of the hind wings.*

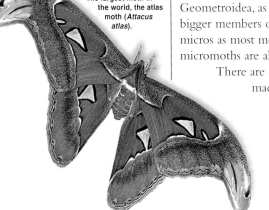

The largest moth in the world, the atlas moth (*Attacus atlas*).

Lepidoptera

Butterflies (*kupu-kupu*) and moths (*rama-rama*) belong to the same insect order, Lepidoptera, coined by the 18th-century Swedish naturalist Linnaeus from two Greek words: *lepis* meaning 'scale', and *pteron* meaning 'wing'. The name derives from the minute scales which cover the wings of these insects. The 'scales' appear as a powdery, dust-like substance when they come into contact with the fingers.

Differences between butterflies and moths

As many butterflies are ornate and colourful, they are traditionally considered more glamorous than moths, most of which are dowdy in appearance. Generally, butterflies are day-flying insects, and when at rest their wings are held erect, while most moths are nocturnal, and at rest their wings are spread horizontally or held roof-like over the abdomen.

The diversity of species in Malaysia

Malaysia has about 1,000 butterfly species; about 90 per cent of those in Peninsular Malaysia are also found in Sabah and Sarawak. The bigger butterflies comprise over 300 species, and belong to the families Papilionidae (swallowtails and birdwings), Pieridae (whites and sulphurs) and Nymphalidae (including danaids, satyrids and amathusiids). The remaining are the smaller but no less beautiful butterflies of the families Lycaenidae (blues and hairstreaks) and Hesperiidae (skippers).

Moths, which are more numerous than butterflies in Malaysia, are commonly divided into the bigger macromoths and the smaller micromoths, based arbitrarily on size. However, some smaller moths, such as Epiplemidae, are considered macros, as they belong to the same superfamily, Geometroidea, as other bigger moths, while some bigger members of the Pyralidae are included as micros as most moths in the family are small. The micromoths are also considered more primitive.

There are about 4,000–5,000 species of macromoths in Malaysia. Common families include Limacodidae (slug moths and nettle caterpillars), Sphingidae (hawk moths and sphinx caterpillars), Saturniidae (emperor moths), Lymantriidae (tussock moths), Arctiidae (footmen, ermine and tiger moths), Noctuidae (owl moths) and Geometridae (looper moths). Most of the 6,000 species of micromoths in Southeast Asia are found in Malaysia. The common families include Pyralidae, Psychidae (bagworms) and Tortricidae (bell moths).

The elegant moon moth (*Actias maenas*), with the longest and most graceful of tails, and its caterpillar on foliage of the *belimbing* tree (*Averrhoa bilimbi*).

Distribution

Half of the butterfly species in Malaysia are found only in the lowlands, below 750 metres above sea level. Those confined to the highlands constitute about one-seventh of the total species. The rest live in habitats on both levels. More than 20 species are

Great mormon (*Papilio memnon*)

Common tiger (*Danaus genutia*)

Malay lacewing (*Cethosia hypsea*)

Blue pansy (*Junonia orithya*)

endemic to Peninsular Malaysia. They include *Euthalia ipona*, *Deramas arshadorum* and *Arhopala cardoni*. Most moths in Malaysia are distributed 500–1000 metres above sea level, the transition zone between lowland and lower montane forest, which contains an overlap of both lowland and montane elements. Most Malaysian butterfly species live in the forest, though these species are low in numbers. The few species with large numbers frequent secondary vegetation.

Some butterfly and moth species are strictly habitat specific, forming an intimate association with that particular ecosystem, for example, the affinity of *Danaus affinis* with mangrove swamp, or the presence of most *Delia* butterflies in highland forest. This specificity could be a result of host–plant specialization, physiological specialization with regard to temperature or other environmental factors, or perhaps even a result of competition.

The species of Mount Kinabalu

Endemics are common on Mount Kinabalu in Sabah; about half of the moth species at 2600 metres and above are only recorded from there. *Diarsia barlowi* and *Phthonoloba titanis* are representative summit examples. It is believed that during the Pleistocene glaciations, when sea levels were lower and dry land connected much of the Sunda Shelf, moths from the Himalayas dispersed to the southeast and colonized Mount Kinabalu. These moths subsequently evolved and radiated from the colonizing stock and became unique species of the mountain.

Sexual dimorphism

In many adult butterflies and moths, the male looks distinctly dissimilar to the female of the same species. For example, the common birdwing (*Troides amphrysus*) has a darker male, but a paler and bigger female, which also has additional spots on the hind wing. The dimorphism should serve as an advantage in mate-finding. Certain species exhibit polymorphism, in which even adults of the same sex may differ morphologically, for example, the great blue mime (*Chilasa paradoxa*).

Sexual dimorphism as seen in *Troides amphrysus*. The smaller male is on the right.

Mimicry and camouflage

Compared to other animals, butterflies and moths appear to be rather delicate and fragile. But the first butterflies or moths are believed to have existed about 100 million years ago, according to fossil evidence which has been discovered in France. One reason for their success in survival is the ability of moths and butterflies to evolve into forms which can protect them from attack. Certain butterflies are poisonous or distasteful to their predators, which

have learnt to avoid them. They develop this defensive ability by the adaptation of their caterpillars to feed on poisonous plants, with the result that their adult form itself becomes poisonous.

Butterflies without such intrinsic qualities nonetheless try to mimic the unpalatable model species externally, in their wing pattern as well as coloration, in order to obtain protection. For example, the forewing of the very large atlas moth looks uncannily like the head of a snake; this undoubtedly helps to frighten off predatory birds. Many caterpillars blend into their natural surroundings extremely well. The caterpillar of a geometrid leaf-feeder, *Hyposidra talaca*, for instance, is brownish and resembles a twig.

Migration

Some butterflies are migratory in habit. For example, in Malaysia the lemon emigrant (*Catopsilia pomona pomona*). This butterfly favours secondary vegetation, and hence is always on the move in search of such pasture. In a migratory flight, many butterflies of the species can be seen instinctively heading towards some predetermined location.

Man–nature connections

Butterflies and moths are important to humans. Most butterflies help to pollinate flowers, as do some moths which visit flowers which only open at night, such as *Micreremites* moths on the flowers of the rattan *Calamus subinermis*.

When caterpillars feed in large numbers on economic crops, they are considered pests. In Malaysia, these include the infestation of crucifers by the diamondback moth (*Plutella xylostella*) as well as the defoliation of *Albizia* trees by the larvae of the grass yellow butterfly (*Eurema hecabe*).

Some of the more flamboyant, but scarce, species are sought after by collectors. Coupled with rapid land conversion, this is threatening their survival. The International Union for Conservation of Nature and Natural Resources (IUCN) lists two Malaysian swallowtail butterflies, *Graphium procles* and *Troides andromache*, as threatened species, while the Rajah Brooke's birdwing is protected by law in Malaysia.

Pingasa ruginaria, a geometrid moth with lichen-like markings on the wings which provide camouflage.

Caterpillar defences

Caterpillars are diverse in shape and structure. Some species have smooth bodies, while others are adorned with filamentous appendages, for example, the tree nymph (*Idea stolli*) (1). Others are armed with thorny spines, for example, the dark red caterpillar of the Malay lacewing (*Cethosia hypsea*) with urticating barbed spines (2). Some, such as the bright yellow lymantriid caterpillar (3), have long urticating hairs to serve as protection.

Attractively packaged mounted butterflies on sale for tourists. In the past, collection of butterflies for such souvenirs was at the expense of butterfly populations in the wild. This has changed with the opening of butterfly parks in Malaysia, which breed the insects. Such parks are proving popular tourist attractions for both locals and foreigners who are able to see species such as the Rajah Brooke's birdwing which previously they could have seen only in the rainforest.

Dragonflies and damselflies

The term 'dragonflies' is often used to refer to both dragonflies and damselflies, which together form the order Odonata; dragonflies form the suborder Anisoptera, and damselflies the suborder Zygoptera. There are at least 220 species of dragonflies and damselflies in Peninsular Malaysia alone. No compilation of Sabah and Sarawak odonates has been made, although 259 species were documented in the whole island of Borneo in 1954.

Cultural differences

In Western folklore, names for dragonflies include dragon's darning needles, able to punish children who misbehave by stitching their lips together while they are asleep.

By contrast, in the East they are a source of food for some communities. Besides Malaysia, other parts of Asia with reports on dragonfly consumption are areas in China, Japan, Laos and Bali. The cooking methods vary, with some communities preferring dry preparations, others moister dishes. A favourite recipe for some is to boil dragonflies in coconut milk with ginger, garlic, shallots and chillies. But dragonflies can also be roasted over charcoal, or steamed in a banana leaf with coconut flesh. Larger species of dragonflies are preferred, and the wings are usually removed before cooking unless they are to be roasted. While such dishes may seem unpalatable to some people, they are relished by others.

History

Fossils of Protodonata which are 300 million years old show that dragonflies were among the largest insects ever to have existed (reaching up to 30 centimetres long and with a wing span of 75 centimetres). Unlike many other animals, dragonflies have not changed their form over time, but they have become smaller, with a wing span of about 20 centimetres. The Odonata owe their successful existence to their unique breeding behaviour and ability to colonize both aquatic and terrestrial habitats, as nymphal forms and adults respectively. They can be valuable to man both aesthetically and scientifically.

Man's fascination with dragonflies goes back to ancient times when Western folklore erroneously regarded them in fear, and thus credited them with evocative names such as horse stingers, flying dragons and devil's darning needles. By contrast, local vernacular names given to these spectacular insects are descriptive in nature, for example, *putir-beliung* ('hurricane'—characteristic of their soaring flight) and *pepatung* ('statue-like'—due to their armoured features).

Behaviour

The damselflies are feeble in flight. They have two pairs of identical wings folded together and held upwards or parallel to the abdomen when at rest. In contrast, the dragonflies have dissimilar fore- and hind-wings which are kept outspread horizontally when perching. Together, 15 families are represented from the 2 suborders, 10 belonging to Zygoptera and 5 to Anisoptera.

Dragonflies are notably light-loving creatures with an affinity for water to display a bustle of diverse activity. Females perch or roost further inland on grasses, shrubs or trees and make only intermittent visits to water to lay their eggs. In contrast, males are resident at water localities and therefore in those areas greatly outnumber the females. Males are conspicuous, perching on prominent places which they use as staging posts to proclaim territories, to intercept females or to combat rival males, or to impale prey on the wing. They enjoy a multitude of insect prey, but are themselves hunted as food by lizards, frogs, spiders, bats and birds.

Odonates and the ecosystem

The aquatic ecosystem determines the richness or poverty of odonate species. An ecosystem with a rich structural complexity provides a variety of niches which help sustain a wider species diversity. Dragonflies take advantage by colonizing the assemblage of available habitats such as fast- or slow-moving streams, quiet pools, ponds and lakes. The expansion of the human population in Malaysia has caused the alteration of natural habitats on which some secretive species depend for existence, and this has resulted in the reduction of these species. However, commoner species, such as *Neurothemis fluctuans*, *N. tulia tulia*, *Orthetrum chrysis*, *O. testaceum*, *O. glaucum* and *Ictinogomphus decoratus melaenops*, have probably benefited from man's activities that resulted in the formation of paddy fields, lakes, ponds and reservoirs. The common riverine species belong to the genera *Vestalis*, *Rhinocypha* and also *Libellago*. To enable conservation measures to be taken, an extensive knowledge of the odonate's natural history is needed.

Perching at a vantage point allows this male *Lathrecista asiatica* to display to conspecific rivals or a mate.

❸

The typical resting posture of a dragonfly with wings outspread.

During resting bouts it is not unusual to see an individual, such as this *Orchithemis pulcherrima*, grooming by rolling its head from side to side.

Dragonflies are beautiful, harmless creatures which play a useful role. The adults are predators of a wide range of insect pests in the paddy fields. The nymphs have a voracious appetite for mosquito larvae and can be used as a biological control agent when placed in water containers that are potential mosquito breeding grounds. Anglers consider the larvae good bait. Dragonflies can reveal information about the extent of man's damage to the quality of freshwater, and a diverse dragonfly population is thus a reliable indicator of a healthy environment.

Dragonfly life by a tropical pond

1. Damselfly
2. Dragonfly
3. Dragonfly laying eggs
4. Damselflies mating
5. Emerging dragonfly adult
6. Damselfly nymph
7. Dragonfly nymph, predatory habit

In contrast to dragonflies, damselflies fold their wings at rest.

A pair of *Cratilla metallica*.

Bees, wasps and hornets

*Malaysia has the greatest diversity (about 35 species) of stingless bees (*Trigona *spp.) in the Asian tropics. The honey of these bees is exploited, as is that of the giant honeybee,* Apis dorsata, *which nests in forest trees, and the common hive bee,* Apis cerana. *Some wasps are also exploited—as biological control agents—while the hornets (large social wasps) can be dangerous to humans.*

The common Malaysian hive bee, *Apis cerana*, which is kept by Malaysian beekeepers for the commercial production of honey

Characteristics of bees, wasps and hornets

Bees (*lebah*), wasps (*penyengat*) and hornets (*tebuan*), which belong to the order Hymenoptera, are noted for their painful stings. People are generally fearful of them, and tend to regard them as pests. The majority, however, are beneficial. Bees are important as pollinators and as a source of honey, royal jelly, bee pollen and wax. Wasps are generally predators and parasites of other insects, and many are useful in the biological control of insect pests. Hornets are large social wasps; in particular, the name is used for members of the genus *Vespa*. They can be dangerous, and readily attack humans who disturb their nests.

It is the group of eusocial (highly social) bees and wasps (including hornets) that are most interesting in terms of their behaviour and social organization. These insects live in colonies comprising hundreds or thousands of individuals. The colony usually consists of a queen (or several queens) and numerous workers; all are females. The workers are female members of the non-reproductive, labouring caste in a colony. Males occur only at certain times of the year, and after mating they are ejected from the parent nest. They are not involved in the establishment of new nests.

Malaysian bees

In Malaysia, the honeybees and the stingless bees belong to the eusocial group of bees. Both honeybees and stingless bees establish new colonies through swarming, but unlike the honeybees, daughter colonies of stingless bees initially maintain contact with the parent nest and depend upon it for food and building materials. The queens of honeybees and stingless bees are long-lived, and the colonies appear to be perennial. Although the workers survive for only a few weeks, the queens live for several years.

The workers and the queens develop from similar eggs, but their development into workers or queens is determined by the quality and quantity of food that the larvae receive from the worker bees. Larvae that develop into queens are regularly fed

Stingless bees

1. A *Trigona* bee on an *Asystasia* flower.

2. The entrance to a *Trigona* nest.

3. A section of the nest interior of a stingless bee, *Trigona thoracica*, showing the queen and workers amongst the brood cells.

4. These remarkable shapes are part of the interior structure of a *Trigona* nest.

royal jelly, a glandular secretion from young workers; the larvae of worker bees are fed mainly with pollen and honey. Among the stingless bees, the brood cells are provisioned with sufficient food for the development of the larvae. The queen lays an egg in each cell before it is sealed up by the workers. The queen's brood cell is larger than workers' brood cells, and is provided with a larger quantity of food.

It has been said that the adult queens of stingless bees depend largely on trophic eggs laid by workers as their source of food. However, some species (such as *Trigona thoracica*) depend mainly on the glandular secretions of young, unpigmented workers, and often pounce on them to solicit such food. It is not yet known whether these secretions are similar to the royal jelly fed to developing queen larvae by young workers of honeybees, or whether they are the elixir of life for the queens of these highly social insects.

Beekeeping in Malaysia

The stinging honeybees, often called simply honeybees, belong to the genus *Apis*, of which Malaysia has at least four species: *A. cerana*, *A. dorsata*, *A. andreniformis* and *A. koschenikovi*.

Gathering wild honey

Honey gatherers go into the jungle in groups of up to 14 people to camp near the tall *tualang* trees (*Koompassia excelsa*) on which the giant honeybee (*Apis dorsata*) builds its exposed nests, often in aggregations.

First they construct a bamboo ladder up the tree. Later, in the darkness of the night, the honey gatherers climb to remove the nests, which are then lowered to the ground in a container attached to a nylon rope. Before removing the nests, a smouldering torch is used to attract the bees, which follow the falling embers to the ground and are then unable to return to their nests in the dark.

The honey, which fetches a higher price than honey from *Apis cerana*, is immediately extracted from the nests using rather crude equipment.

A bee spirit (*hantu lebah*) of the Jahut people from whom permission must be obtained before honey is removed from the *tualang* tree.

The common hive bee is *A. cerana*, which is slightly smaller than the familiar *A. mellifera* of the temperate and subtropical countries, and has a much lower yield of honey. There are more than 1,000 registered beekeepers in the country, maintaining more than 5,000 hives of *A. cerana*. In Malaysia, beekeeping is still in its infancy, and the Malaysian Bee Research and Development Team (based at Universiti Putra Malaysia) is trying to promote and to improve beekeeping with indigenous bees.

The giant honeybee (*Apis dorsata*) builds exposed nests, often in clusters, high on tall *tualang* trees (*Koompassia excelsa*) in the forests. Honey gatherers camp near the trees for the night-time collection of the hives. The honey of the giant honeybee fetches a higher price than the honey of *A. cerana* reared in box hives.

The stingless bees, belonging to the genus *Trigona*, also produce honey, and their nests are occasionally exploited for their honey and resin. These bees occur only in the tropical and subtropical regions of the world.

Malaysian wasps

The majority of wasps are solitary in habit, and include numerous parasitic as well as predatory species. Many are tiny insects, some less than 0.5 millimetre in length. However, *Scolia procer*, a parasite of the rhinoceros beetle grub, is a large wasp measuring up to 55 millimetres in length. The solitary pompilid wasp hunts spiders, which it stings and then drags along into a nest. An egg is then laid on the prey; this will later be fed upon by the larva. The social wasps are all predacious, and belong to the family Vespidae. Their social organization ranges from that of the primitively social to the eusocial type, represented by the hornets.

Malaysian hornets

In temperate and subtropical countries, a hornet colony is founded by a single queen. However, *Vespa affinis* and *V. tropica*, common hornets in Malaysia, establish their colonies through the cooperative

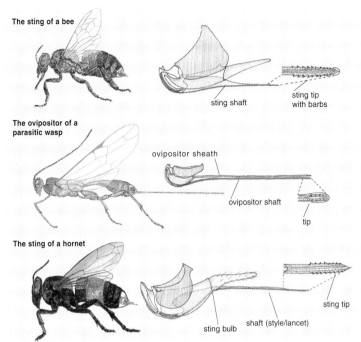

The sting of a bee

The ovipositor of a parasitic wasp

sting shaft
sting tip with barbs

ovipositor sheath
ovipositor shaft
tip

The sting of a hornet

sting bulb
shaft (style/lancet)
sting tip

The sting of the bee, wasp and hornet

The sting of a bee or wasp is a modified ovipositor (egg-laying apparatus) of a female insect. Hence, only females of these insects can sting.

The bee feeds on pollen and nectar, and uses its sting solely for defence. This often results in the sacrificial death of the defender as the bee is unable to withdraw the sting without tearing out part of the viscera from her body due to the very sharp barbs at the tip of the sting.

The parasitic wasp uses her ovipositor to pierce through host tissue to lay eggs or to enable the body fluids on which it feeds to ooze out. The predatory wasp uses her sting to paralyse the prey, which may be another insect or a spider. She then lays an egg on or beside the paralysed prey so that the newly hatched young will have a fresh but inactive prey to feed upon.

The hornet uses her sting for both defence and offence, and the larva feeds on masticated food prepared by the adult.

efforts of several queens. The nests of *V. affinis* are suspended from branches of trees or from the eaves of buildings, while those of *V. tropica* are found in concealed locations. The two species are superficially alike in having black and yellow bands on the abdomen. However, *V. tropica* is distinctly larger, and the yellow marking is confined to the second abdominal segment, whereas in *V. affinis* the yellow band occurs on both the first and second abdominal segments.

Unlike honeybees, nest-founding among the hornets does not involve the worker caste. A colony is initially founded by a single queen, who may be joined by other queens. During the early stage of nest-founding, the hornet queen is involved in egg-laying, foraging and nest construction, as well as in feeding the young. Cooperative nest-founding by multiple queens helps to ensure the success of a new colony. When one or more queens are out foraging, there is at least one guarding the young nest against attack by predators.

The first to develop among the progeny are the workers, which are derived from fertilized eggs. They relieve the queen of her various activities, except egg-laying. During the later stages, male and female reproductives (potential queens) begin to appear; eventually, the colony starts to decline. The female reproductives originate from fertilized eggs, just like the workers, whereas males originate from unfertilized eggs. Hornet colonies do not last for more than a year.

A colony of wasps
(*Liostenogaster flavolineata*)

Vespa affinis indonensis
MALAYSIA 20¢

1. The common Malaysian hornet, *Vespa affinis*.

2. A young nest of *Vespa affinis* with multiple queens.

3. A postage stamp showing the *Vespa affinis* wasp, one of four stamps depicting Malaysian wasps issued by Pos Malaysia in 1991.

Ants and termites

Malaysia is home to several hundred of the 9,000–10,000 ant species found throughout the world, some well known for the ferocity of their sting. Of the 2,300 species of termites in the world, nearly 200 have been recorded in Malaysia. These include serious attackers of plantation crops as well as those which destroy wooden structures.

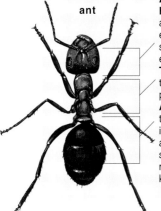

Giant forest ants (*Camponotus gigas*) and sap-sucking bugs have a mutualistic relationship. The bugs provide food in the form of honeydew excreta for the ants and, in return, the ants protect them from predators.

Brown crazy ants (*Anoplolepis longipes*) 'milking' an adult membracid.

Differences between ants and termites

Ants (*semut*) and termites (*anai-anai*) are both social insects. They live and work together as a family, called a colony. Although they look similar, their body structure and genetics are totally different. Ants have three clearly defined body parts; termites have only two. Termites, which belong to the order Isoptera, are believed to be evolved from cockroaches. Ants, from the order Hymenoptera, are related to wasps.

Confusion between ants and termites often arises because termites are commonly known as 'white ants', and also because of their identical foraging pattern and behaviour. While foraging, they release volatile chemicals known as pheromones to mark their trail. This is the reason why ants and termites are always seen moving in an organized pattern.

Distribution

Ants are found everywhere except the North and South Poles. In Malaysia, those often seen in gardens are the weaver ant (*Oecophylla smaragdina*), the fire ant (*Solenopsis geminata*) and the brown crazy ant

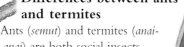

Nest of the weaver ant (*Oecophylla smaragdina*).

Weaver ants at work in nest construction. The ants are pulling two leaf edges together and holding them in place, to be bound with larval silk.

(*Anoplolepis longipes*). The giant forest ant (*Camponotus gigas*) is conspicuous in the forests, while the black crazy ant (*Paratrechina longicornis*) is common in kitchens, where it is attracted to sugar and other foodstuffs.

Termites are found only in tropical and subtropical regions. In Malaysia, the common species include *Coptotermes curvignathus* (subterranean species), *Macrotermes gilvus* (mound-building species) and *Dicuspiditermes nemorosus* (soil pillar species).

Ant and termite body structure

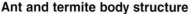

Head: Includes eyes, mandibles and antennae. Unlike ants, some termites are eyeless, and the antennae are made up of small bead-like segments instead of being elbow-shaped.

Thorax: In both groups this supports the three pairs of legs, as well as wings, if present.

Abdomen: In the ant this is connected to the thorax by a thin waist, the petiole, which is absent in the termite. The oval part of the ant's abdomen—the gaster—may have a sting, while the tenth segment of the termite may bear a pair of sensory appendages known as cerci.

Garden ants

Brown crazy ants are yellowish brown insects about 3–4 millimetres long often found in orchards and plantations, and sometimes also in houses. They make their nest under leaf litter or debris on the ground. A harmless species, they run aimlessly when molested, hence their name.

Brown crazy ants (*Anoplolepis longipes*)

Weaver ants (*kerengga*) are relatively large, up to 8 millimetres in length. The worker ants are orange or rusty red, while the queens are green (an unusual colour among ants) and the males black. They do not sting, but inflict painful bites with their formidable mandibles. They also release formic acid from the end of the gaster, which can cause irritation to human skin.

Fire ants (*semut api*) are reddish yellow insects 3–4.5 millimetres long. The soldier caste, with a distinctly enlarged head, is bigger, about 8 millimetres in length. Fire ants are known for the painful burning sensation of their bite.

The importance of ants

Many people regard ants as pests as they damage crops, and some inflict painful bites and stings. However, these creatures are important in many aspects. They are a source of food for other animals, and their great abundance makes them effective as biological control agents. The Malaysian cocoa black ant (*Dolichoderus thoracicus*) secretes volatile phenolic substances from the gaster, which frighten intruders away. They control the mirid bug (*Helopeltis theobromae*), which is a pest in cocoa plantations. Efforts have also been made in Sabah to use these ants to control the lepidopteran cocoa pod borer (*Conopomorpha cramerella*). Terrestrial ants such as *Odontoponera transversa* and *Diacamma rugosum* help to decompose organic matter, recycle nutrients and enrich soil. In short, ants are also friends of man!

The termite threat

Termites are a greater threat to humans because they not only attack timber and wood products, but also living trees, including plantation crops such as rubber, coffee and cocoa. Every year throughout the world, property and amenities worth nearly US$22 billion are lost due to termite attacks. In Malaysia, the most serious damage is caused by the termite genus *Coptotermes*, which, besides attacking agricultural crops, also causes damage to furniture and structural timbers of buildings. The termite method of attack is to remove all palatable wood except the outer layers, which provide them with shelter and protection. Their actions are not wholly destructive, however. In nature, termites play a role in recycling dead woody materials and so contribute to soil formation.

Ant and termite architecture

Ants are hard-working and creative. Some heart–gaster ants (*Crematogaster* spp.) construct carton nests from half-digested wood, leaves and soil particles. The weaver ants build their nests from living twigs and leaves, which are held together by larval silk. Some ants live inside plant structures. Domestic ants, such as the pharaoh's ant (*Monomorium pharaonis*), occupy cracks in the wall, spaces in

piles of discarded clothing and in between leaves of books, and any other places which are dry and fairly warm. Inside ant and termite nests, multiple chambers connected by tunnels are used for specific purposes, such as a nursery, queen's chamber, storage chamber and even a sewage chamber.

Termites are also great architects, as can be seen from the various forms of nest structures. The wood-feeding termites, such as the genus *Microcerotermes*, make their nests in the form of a honeycomb of small cells with walls of carton material derived from partly digested wood, and rich in the rigid plant polymer, lignin. The outer layer is often protected with soil mixed with carton, shaped in a design, such as a mushroom or umbrella, to protect the nest from the rain or bright sunshine.

The soil-feeding termites, *Dicuspiditermes*, use their excreta as a building material, composed mainly of fine clay with vegetable residues. The fungus-feeding termites build their nest with sand particles cemented with fine clay and saliva. Termites of the genus *Macrotermes* can build a mound up to a few metres high. Some termites, such as *Nasutitermes*, build nests high up in trees. Perhaps this keeps them safe from predators. Others, such as *Coptotermes*, keep the entire colony underground without any mound.

Termite mounds

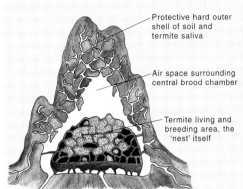

Protective hard outer shell of soil and termite saliva

Air space surrounding central brood chamber

Termite living and breeding area, the 'nest' itself

Various species of termites make mounds of different shapes, as seen in these examples.

A Jahut carving of a termite mound spirit that is believed to bite the legs of anyone who goes near it.

Crematogaster inflata is a lowland rainforest ant species with a striking yellow metapleural gland at the back of its thorax which looks like a miniature life jacket. The gland produces whitish secretions which immobilize the ant's prey.

Myrmecodia tuberosus give ants shelter in their tubers, and in return absorb nutrients from the ant nest.

Termites

1. A *Microcerotermes dubius* nest on a tree trunk. This termite species is a serious pest in forest plantations because it attacks living trees.

2. The termite queen (*Microcerotermes dubius*) surrounded by workers. The queen, with an enlarged abdomen, is merely a vast egg-laying machine with a single ovary containing more than 2,000 strings of developing eggs.

3. The internal damage to a log caused by wood-feeding termites.

4. Some termites, such as *Nasutitermes*, have a pear-shaped head with a nozzle which ejects defensive fluids.

Stick and leaf insects

Of the more than 2,500 species of stick and leaf insects recorded worldwide (with many more awaiting discovery), probably 200–250 species of these amazing insects inhabit the rainforests of Malaysia. They include the stick insect Phobaeticus kirbyi, *renowned as the longest insect in the world (up to 30 centimetres) and the most famous of phasmids, the leaf insects of the genus* Phyllium, *sometimes called 'moving leaves'.*

Phobaeticus serratipes, a relative of the longest insect of all.

The 'moving leaves' of the genus *Phyllium* are the most remarkable leaf-mimicking insects known. *Phyllium giganteum*, shown here, was only recently documented scientifically, and is the largest species of the genus. It appears to be very host-specific, feeding on leaves of the riverside *Saraca thaipingensis*.

Bright colours displayed by some phasmids (in this case, *Tagesoidea nigrofasciata*) serve to break up the shape of the insect within green foliage, and so fool potential predators.

Stick and leaf insects

Stick and leaf insects (*belalang ranting* and *belalang daun*), for a long time included as a family of the order Orthoptera together with grasshoppers and praying mantids, are now considered a separate order, Phasmida. They are primarily tropical in distribution. Species found in the montane forest are usually smaller than those from the lowlands. In general, males are markedly smaller and more slender than females; sometimes the difference is quite significant. Many species are parthenogenetic, that is, females are able to reproduce without mating.

The most famous of the phasmids are the leaf insects of the genus *Phyllium*. The females, in particular, can be so well camouflaged that they are almost impossible for the untrained eye to detect. Their green bodies are dorso-ventrally compressed and the wing covers have vein-like patterns, complete with a 'midrib' and 'fungal spots' that one expects to see on leaves in the forest. The legs have broad expansions to further mimic plant structure.

The antennae are short in the females and long in the winged males. In captivity, these leaf insects have been observed nibbling at each other, mistaking one another for leaves!

Generalist and specific diets

All phasmids are phytophagous, that is, they feed on plants. Although many are generalist in their dietary requirements, some are host-specific. The stick insect *Lonchodes harmanii* from Mount Kinabalu, for example, has only been found feeding on the mountain mussaenda (*Mussaenda frondosa*) while *Carausius sanguinoligatus* and *Neachus redtenbacheri* from the same mountain feed exclusively on a wild raspberry (*Rubus moluccana*) and a *Macaranga* sp. respectively. *Phyllium giganteum* appears to feed only on the foliage of *Saraca thaipingensis*.

Endemism and distribution

Because of their rather specific food requirements and low densities, many phasmids are endemic or specific to certain areas, such as Borneo or parts of it. *Haaniella echinata* is an example of a relatively common species found only in the lowlands of Sabah, northern Sarawak and Brunei. *Haaniella saussurei* is typically a Sarawak species and has not been recorded from Sabah.

In Peninsular Malaysia, *Prisomera repudiosa*, a very spiny, wingless species, appears to be endemic to the foothills of the Cameron Highlands. Two of the three known species of *Gargantoidea* — *G. triumphalis* and *G. tesselata*—are also restricted to Peninsular Malaysia. However, when more field studies are carried out, it is expected that some of the little-known, so-called endemic species will be found to have a wider distribution than previously thought.

Artful defence: Mimicry and camouflage

Remaining absolutely motionless is the primary defensive behaviour of phasmids. Some stick insect species, such as *Asceles margaritatus* and many species of the subfamily Necroscinae, sway their body sideways to simulate movements of twigs in

The nymph of the leaf insect *Phyllium bioculatum* is red when first hatched, so it is not noticeable among the forest litter. It changes to green as soon as it starts feeding on plants, so that it resembles a leaf.

a gentle breeze. Most species of phasmids have a superb 'cryptic' coloration that enables them to pass unnoticed even when they feed in the open. *Dares validispinus*, a forest floor stick insect found throughout lowland Borneo, has a patterning that is

Lonchodes hosei is just one of the species which are perfect mimics of twigs. When remaining still they are almost undetectable.

Haaniella echinata, one of the most common lowland rainforest stick insects endemic to Sabah and Sarawak, has a body length of 11.5 centimetres and has underdeveloped wings. When threatened by a predator, it stridulates—rubs its wing covers violently against its lower wings—to produce a hissing sound. This stick insect lays the largest eggs known in the insect world, up to 1 centimetre long! Although covered with spines, it is eaten by a species of monitor lizard in Sarawak.

impossible to detect on the forest floor. The female of the gigantic stick insect *Heteropteryx dilatata* mimics a bright, young, yellow-green leaf. The male, on the contrary, is brown and resembles a twig. This species, together with *Haaniella echinata*, reacts fiercely when predators approach, stridulating to produce a hissing sound akin to that of a snake.

When they sense danger, some stick insects, particularly of the subfamily Necroscinae, emit a fluid that can smell so strongly that even humans are repelled. *Trigonophasma* spp. from the lowlands are good examples of species which do this. The mountain *Necicroa terminalis*, likewise, secretes a brown, smelly fluid when handled, leaving a stain on the hand for more than 24 hours.

Sexual dimorphism

Size and form differences between the sexes are particularly marked in phasmids. Males are significantly smaller and more slender than females, and often have functional wings. The wings and lighter weight of males allow them to fly in search of mates. This is significant in the leaf insect genus *Phyllium*, where females are often found high up in trees and do not appear to descend to the ground, while males fly around to look for females.

The stick insect *Phenacephorus cornu-cervi* from Mount Kinabalu is an excellent example of extreme sexual dimorphism in phasmids. In the female of this species, there are extensive leaf-like outgrowths on the body, and similar expansions on the legs which can change colour according to the surroundings. The male totally lacks these ornamentations and is a very slender, plain insect by comparison. In the Borneo endemic *Galactea*, the wingless female is armed with two rows of spines along the bright green body, while the brown, winged male is almost devoid of spines. The only feature shared by the two sexes is a pair of spines on the head! Sexual differences can be so extreme in phasmids that entomologists have sometimes described each sex of the same species as different species, sometimes even assigning them to different genera!

Eggs and egg-laying

Phasmids often lay their eggs one at a time. The eggs have a remarkable resemblance to the seeds of rainforest plants. The largest insect eggs known belong to *Haaniella echinata*, a common lowland rainforest stick insect. Egg-laying behaviour among stick insects is varied. *Heteropteryx dilatata* almost always lays its eggs on rainy nights when the soil is soft and soggy, using its strong, rigid ovipositor to excavate a pit large enough to bury up to 10 eggs, before ascending to nearby shrubs. *Haaniella echinata* simply drops her large peanut-size eggs onto the forest floor; these eggs may take up to a year to hatch. The Necroscinine genus *Trigonophasma* lays eggs in clutches of 6–8, gluing them on the lower sides of leaves or bark. The most amazing eggs are perhaps those of the genus *Asceles*. The eggs of this genus are armed with a sharp pin on one end, and the female lays the eggs by piercing the pins through a leaf, to which they remain very firmly attached. *Asceles margaritatus*, from the lowland rainforest in eastern Sabah, has been observed to deposit up to three eggs on a single leaf.

Natural enemies

The eggs of phasmids are often eaten by ants or other predatory animals. Birds generally relish fleshy insects such as phasmids. *Haaniella echinata* is apparently the staple food of a monitor lizard (*Varanus* sp.) in Sarawak. Adult phasmids, as well as the thorny penultimate nymphs of *Haaniella echinata*, are commonly preyed upon by spiders. Parasitism by flies or wasps is also not uncommon among adults, while mites are common as ectoparasites, in particular in the genus *Haaniella*, where they are found in the abdominal pits.

The male (blue), female and nymph of the stick insect *Marmessoidea rosea*. Both male and female have red hind wings. The bright colour of this species is a warning sign; it is able to spray a very strong-smelling substance when handled.

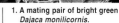

1. A mating pair of bright green *Dajaca monilicornis*.

2. *Phaenopharos struthioneus*, with small red wimgs which it opens when molested, distracting the predator long enough to escape.

The thorny tree-nymph stick insect

One of the best known of Malaysia's stick insects, *Heteropteryx dilatata*, is not at all stick-like. The light green female, which looks like a leaf, is well armed with sharp spines, and is certainly difficult for any predator to make a meal of. The drab brown male, which is much more stick-like, has fully functional wings, in contrast to the female's vestigial wings. The male and female differ not only in looks, but in defence strategy. Like many other phasmids, the male drops to the forest floor, remaining there motionless until danger has passed. In contrast, the female clings on with her spiny hind legs, at the same time producing a snake-like hissing sound by rubbing the elytra and the hind wings together.

In the 1950s, this insect was kept by many people in Pulau Pinang who believed the droppings of this species, taken in a glass of hot water, had medicinal properties, curing diarrhoea and gastroenteritis. The 'secret' to this belief was probably the diet of the stick insects as they fed on leaves of the guava tree, which is known to have medicinal properties.

Heteropteryx dilatata has been collected in large numbers by the Orang Asli, not only for butterfly farms and souvenir shops but also for agents who export both the live insects and their eggs to Europe.

Beetles and weevils

Malaysia's rainforests support a very high diversity of beetle species. A study of the leaf beetles of the subfamily Galerucinae from the Danum Valley, Sabah, revealed that an area of about 2 square kilometres contained about 200 species. While many of Malaysia's beetles cause damage to agricultural crops, there are others which prey on other harmful insects.

Head of a beetle and weevil

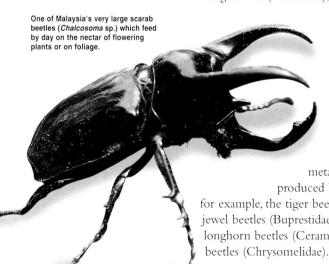

antenna

beetle

biting jaws

rostrum

clubbed antenna

eye

weevil

Beetles and weevils belong to the same order, but weevils differ in having an elongated head (rostrum). Their clubbed antennae are attached to a depression on each side of the head.

One of Malaysia's very large scarab beetles (*Chalcosoma* sp.) which feed by day on the nectar of flowering plants or on foliage.

Characteristics

Beetles are most easily distinguished from other insects by their hard front wings (elytra) which cover the membranous hind wings and abdomen when the beetle is not in flight. The elytra, however, are not used during flight.

Their mouthparts are of the biting type, unlike bugs which suck. Weevils are also beetles, but differ in having an elongated head (rostrum) with clubbed antennae attached to a depression on each side. The various forms, colours and modes of life of the numerous beetles are fascinating.

Beetles form the order Coleoptera, the largest order of insects, with over 350,000 species worldwide. Weevils form the Curculionidae, the largest of the families of the Coleoptera. Beetles are found in almost every habitat, terrestrial as well as aquatic. Their modes of life are varied: many are phytophagous, many are predaceous, some are scavengers, others feed on fungi, and a few are parasitic. One of the biggest beetles in Malaysia is the atlas beetle (*Chalcosoma atlas*) (Scarabaeidae), with a body length of about 130 millimetres. In comparison, the length of the rice weevil (*Sitophilus oryzae*) is about 3 millimetres, and the rhinoceros beetle (*Oryctes rhinoceros*) about 40 millimetres.

Beetles display a wide variety of colours. Coloration is due to the presence of melanins (brown or black) and carotenoids (red, orange or yellow), or an optical effect produced by special cuticular structures. Some of Malaysia's beetles, such as the stag beetles (Lucanidae), the scarab beetles (Scarabaeidae) and the weevils (Curculionidae), are beautiful creatures with their shiny black or orange coloration. But the most beautiful are those with brilliant metallic coloration produced by an optical effect, for example, the tiger beetles (Cicindelidae), the jewel beetles (Buprestidae), and some of the longhorn beetles (Cerambycidae) and leaf beetles (Chrysomelidae).

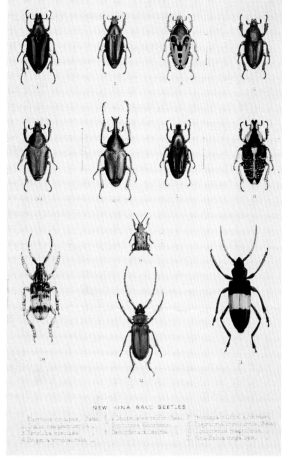

NEW KINA BALU BEETLES

A selection of beetles found on Mount Kinabalu in Sabah by 19th-century naturalists, reproduced from John Whitehead's book, *Exploration of Mount Kina Balu*, first published in 1893.

Life cycle

Beetles undergo complete metamorphosis, the life cycle moving from egg to larva to pupa to adult. The *kedondong* beetle (*Podontia quatuordecimpunctata*), amongst the earliest of Malaysian beetles to be studied locally, can be recognized from its orange body, with 14 black oval spots on the elytra. The female deposits her eggs in batches, the number ranging from 18 to 59. The incubation period for the eggs is about 6–8 days followed by a larval stage of about 11–18 days. When fully grown, the larva enters the soil, encloses itself with an oval cocoon made from particles of earth within which it pupates. It takes between 14 and 29 days for the adult to emerge from the soil.

How beetles fly

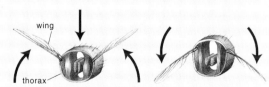

wing

thorax

Although beetles have two sets of wings, it is only the membranous hind wings which are used for flight.

The power needed by beetles to flap their wings comes from muscles in the thorax. As these muscles contract, the thorax expands and contracts, forcing the wings up and down.

The direction of flight is determined by muscles at the base of the beetle's wings.

Sexual dimorphism

In beetles, sexual dimorphism (differences between males and females) takes a number of different forms. These include enlarged mandibles in stag beetles, a depressed or sunken face and broadened maxillary palpi in leaf beetles, serrated antennae in click-beetles (Elateridae) and horn-like processes on the pronotum in scarab beetles (Scarabaeidae).

In one extreme instance of sexual dimorphism, the female is of very different form. The adult female of *Duliticola* spp., a net-winged beetle (Lycidae), often known as 'trilobite larvae', is very large (50 millimetres), is wingless and has a similar form to larva. The adult male, in contrast, is much smaller (10 millimetres) and winged.

Female net-wing beetles ('trilobite larvae') (*Duliticola belladonna*), an extreme example of sexual dimorphism.

Polymorphism

Polymorphism (multiple forms) in beetles is most obviously displayed by the various colour patterns of the ladybirds which feed on aphids. The pronotum and elytra of the beetles have brown or black spots on a yellow, orange or red background. The absence of spots and bars or their coalescense gives rise to polymorphism. *Coelophora inaequalis* has nine polymorphic forms, from a spotless to an almost completely black form, while in *Harmonia arcuata* the completely black form is absent.

The tortoise beetle (*Aspidomorpha miliaris*) has black spots at the four outer angles of the thin elytra. The larvae of this beetle have soft spines along the body.

Pest beetles

A few beetles are pests that cause serious damage to agricultural crops. A brown hairy ladybird (*Epilachna sparsa*), with a number of black spots on the elytra, is destructive to cultivated plants, such as brinjals, tomatoes, cucumbers and watermelon. Mango growers often find pieces of young leaves all over the ground early in the morning, a result of the female leaf-cutting weevil (*Deporaus marginatus*) cutting off the leaf after eggs have been laid on it. The emergence of the adult weevil usually synchronizes with new flushes of young leaves. The rhinoceros beetle is a major pest of coconut and oil palm. Adults feed on succulent palm spears, while the larvae feed on dead organic materials, such as rotting palm stems and compost heaps. Another major pest of coconut is the red stripe weevil (*Rhyncophorus schach*), the larvae of which feed on the crown and trunk of the palms. Stored products such as rice are commonly attacked by the rice weevil.

Beneficial beetles

Many families of beetles are predators, devouring other insects. For example, a number of ladybirds (Coccinellidae), ground beetles (Carabidae) and rove beetles (Staphylinidae) feed on brown planthoppers

Diverse beetle species

1. *Cyriopalus wallacei*, one of the longhorn beetles named for their distinctive antennae.

2. *Batocera albofasciata*, a longhorn beetle which is able to make shrill noises, possibly to deter predators; its larvae attack tree trunks.

3. *Lepidiota stigma*, a cockchafer beetle which attacks the roots of agricultural crops.

4. The leaf-cutting weevil (*Deporaus marginatus*), which lays its eggs on mango leaves and then cuts them off.

5. The two-horn rhinocereos beetle (*Xylotrupes gideon*), a pest of coconut and oil palm plantations, where it feeds on young palms.

6. The rhinoceros beetle (*Oryctes rhinoceros*), an agricultural pest as it feeds on the sap of palms.

(*Nilaparvata lugens*), a serious rice pest. Dung beetles (Scarabaeidae) play an important role in the conversion of livestock excrement to humus, and also improve the fertility of the soil by their burrowing activities. A large number of beetles frequent flowers, and thus play the role of pollinator.

Introduced beetles

Introduced beetles which have become part of Malaysia's beetle fauna include two beneficial species. In 1981, the African oil palm weevil (*Elaeidobius kamerunicus*) was released in Malaysian oil palm plantations for pollination of the palms, previously done mainly by plantation workers. Due to the tremendous increase in yield, the beetle was dubbed 'the million ringgit weevil'. Another introduced beneficial beetle, the leaf beetle (*Metrogaleruca obscura*) from South America, is used in Selangor coconut plantations to control the woody weed *Cordia curassavica*.

Dung beetles

Dung beetles (Scarabaeidae) play an important ecological role by utilizing the dung of animals as food for their young. In this way, the soil is enriched both by the animal dung and by the burrowing of the dung beetles.

The beetles roll pieces of animal dung into balls, which are pushed down a tunnel to underground chambers excavated by the beetles. Here, the female lays eggs. Once an egg hatches, the larva burrows its way into the centre of the dung ball, where it feeds on the dung, and pupates there. When mature, the beetle bites its way out of the ball.

Fireflies

Fireflies are beetles belonging to the family Lampyridae, which shine in the night like fire to attract sexual partners. Light is emitted from luminous organs located on the posterior abdomen, produced by the oxidation of luciferin pigments in the presence of an enzyme, a biochemical reaction which produces light but not heat.

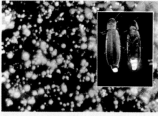

It is the folded-wing fireflies (*Pteroptryx* spp.) which have developed the striking synchronous flash pattern. The males of each species flash in unison, but each species has a different pattern. Lack of synchronization may indicate the presence of more than one species in the same tree or the presence of females, which do not synchronize.

In Malaysia, the most popular site to observe this remarkable phenomenon is in Kuala Selangor, on the shores of rivers in mangrove forests. Here, thousands of fireflies may group together on a tree, beginning their display about one hour after sunset. A really bright display lasts for about 2–3 hours, continuing at lower levels until dawn.

Cicadas and bugs

Cicadas and bugs, which are very common insects in Malaysia, belong to the orders Homoptera and Hemiptera respectively. People often apply the term 'bug' to any kind of insect, but only members of the order Hemiptera are true bugs. Cicadas and bugs have similar piercing and sucking feeding habits which are different from those of other insects, and it is these habits which make them agricultural pests.

As plant eaters, cicadas do not eat other insects; however, they are preyed upon, as by this cricket (1). Although quite large insects, cicadas are not easily seen against the tree trunks or branches to which they cling, as these two Malaysian cicadas, *Cosmopsairria latilinea* (2) and *Dundubia* sp. (3).

'Raining' trees

As cicadas feed on plant sap, which is mostly water, they must suck a large volume of sap to obtain sufficient nutrients. The excess water is excreted sporadically in a shower of fine droplets. Thus, when there are a large number of cicadas in a tree, it can appear that a shower of rain is emanating from the tree. This can present a puzzle to the recipients of such a shower as it is obviously not raining, but yet the source of the shower is very elusive, cicadas not being easily sighted as they cling to a branch.

Homoptera and Hemiptera

The clue to the distinction between Homoptera and Hemiptera lies in their names. Common to both is the second half, 'ptera', which comes from the Greek *pteron*, meaning 'wing'. *Homos*, 'the same', identifies the Homoptera as insects whose front wings are an even texture all over. *Hemi*, 'half', reflects the division of the front wings of Hemiptera into two areas—tough at the base and soft on top.

Homoptera and Hemiptera differ from other insects in their feeding habit. While other insects bite, the Homoptera and Hemiptera suck up their food. Thus, their mouthparts are different in form from those of other insects.

The mouthparts of both cicadas and bugs, called a beak or rostrum, are formed for piercing and sucking. The piercing organs consist of four long mandibular and maxillary setae, and the whole structure is enclosed in a jointed sheath called the labium. To obtain their food, cicadas and bugs puncture the young woody growth of plants and suck out the sap, a habit that makes bugs plant pests.

Both Homoptera and Hemiptera undergo simple metamorphosis during post-embryonic development; that is, the young (nymphs) and adults are very similar, except in size. Wings appear as bud-like outgrowths in the early instars and increase in size until the last instar.

Cicadas

Cicadas (*riang-riang*) belong to the family Cicadidae. They are generally large insects, ranging from 2 to 10 centimetres in length, but are difficult to see as they sit very still on trees and bushes. Cicadas have a stout body and a wide, blunt head with bristle-like antennae and large eyes pointing outwards. The two pairs of transparent wings are membranous, with a mass of several veins along the costal margin and also cross-veins as branches of the primary veins.

The most distinctive feature of cicadas is the musical organs of the males, which produce the cicada 'songs' heard in the forest. Each species has its own characteristic sound, the purpose of which is to both attract the female and to mark out territory. When attacked, cicadas produce a characteristic sound, which often scares off an attacker.

The cicada's 'song'

Male cicadas are known for the 'songs' they use to attract females and to mark their territory.

The stridulating (sound-producing) organs are located at the base of the abdomen of male cicadas. A pair of timbals, which resemble the tightly drawn skin of a hollow drum, have muscles attached to the underside which allow the timbals to click in and out at high speed. The cicada's song is produced by the vibration of the timbals, and is transmitted by the air contained in the large air sacs which, together with the body tension, also control the volume and quality of the song.

The structure of the sound-producing organs varies in different species of cicadas. The song, too, varies from species to species so that a mate can attract a female of his own species.

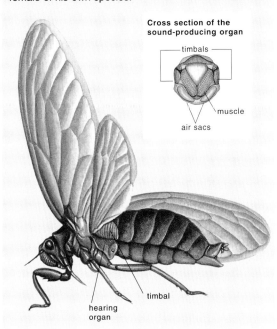

Cross section of the sound-producing organ

timbals

muscle

air sacs

timbal

hearing organ

Female cicadas lay large numbers of eggs inside twigs which they have split open with their ovipositor. After they hatch, the nymphs burrow down into the soil, where they feed on the sap of plant roots. When the cicadas mature, a process which may take several years (the length of time varies from species to species, but can be as long as 17 years), they climb up a tree and emerge from the nymphal case. The long maturation period of cicadas has caused them to be regarded as symbols of eternal youth or immortality.

Cicada development

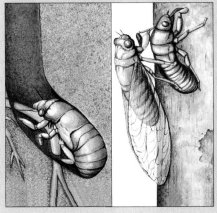

The female cicada uses her ovipositor to cut slits in twigs, in which she then lays her eggs. When the nymphs hatch, they drop to the ground and burrow into the soil, where they feed on the sap of tree roots (left). Here they slowly mature, a process that may take many years to complete. The mature insects climb up the tree trunk, where they emerge from the nymphal case (right).

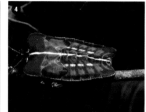

The numerous fascinating bugs found in Malaysia include cotton stainers (*Dysdercus cingulatus*) (1), the man-faced bug (*Catacanthus incarnatus*) (2), the shield bug (Pentatomidae), a juvenile of which is shown here (3), and the stink bug (*Pycanum rubens*) (4)—the illustration here is a nymph of this species.

Bugs

Many species of bugs eat vegetation. Among them are, in Malaysia as in the rest of the world, some of the most damaging pests of cultivated plants. On the other hand, some species play a beneficial role by feeding on insect pests. Others, such as the Indian lac insect (*Laccifer lacca*), are of commercial importance. The amount of wax or lac exuded by the females is so great that twigs are coated with it. The twigs are cut, the wax melted and refined, and used in shellac production.

Defence mechanisms of bugs

Bugs all have a means of defending themselves against predators. Taste, colour, and camouflage are just a few of the techniques used as a means of defence. Some jump powerfully when faced with danger; others taste foul. Bright colours are usually a warning sign that those bugs are distasteful. Stink bugs emit a pungent smelling substance, either squirting or smearing this on their enemy before making a quick escape. Other bugs, because of their shape or colour, blend in perfectly with their environment, avoiding unwanted attention. Spines or a tough skin make some bugs difficult to eat, forcing predators to look for easier prey. Some bugs deter possible attackers by the exudation of wax or foam, while there are also bugs which mimic other insects, such as fierce ants or wasps, to avoid attack.

BUGS AS AGRICULTURAL PESTS		
SCIENTIFIC NAME	COMMON NAME	CROP
Nephotettix spp.	Green planthopper	rice
Nilaparvata lugens	Brown planthopper	rice
Sogatella furcifera	White-back planthopper	rice
Scotinophara coarctata	Black rice bug	rice
Leptocorisa spp.	Rice ear bug	rice
Nezara viridula	Green rice bug	rice
Helopeltis theobromae	Mosquito bug	cocoa
Rhynchocoris humeralis	Green citrus bug	citrus
Idioscopus spp.	Leafhopper	mango
Mictis longicornis	Stink bug	lady's finger
Dysdercus cingulatus	Fruit bug	lady's finger
Aphis spp.	Aphid	vegetables

Lantern flies

Next to the Cicadidae, the family Fulgoridae possesses some of the most uniquely formed and most splendidly coloured of known Homoptera. Lantern flies often have an elaborate pattern of many colours—red, green, black, blue, and yellow. They are divided into several subfamilies. The Lanternariinae are characterized by the most remarkable prolongations of the head. *Fulgora* is its largest and best known genus. Its members have long been known by the common name of 'lantern flies', based originally on the belief that the apex of the cephalic process, normally reddish, was phosphorescent.

Several species of *Fulgora* can be found in Malaysia, for example, *F. pyrorhyncha*, *F. spinolae* and *F. occulata*. The unique characteristic of this insect is the cephalic process or rostrum, which is nearly as long as its body. The purpose of this rostrum is probably sensory and balance-related. In Malaysia, these insects can be found in rainforests such as in Taman Negara, and in fruit orchards. Normally they are found on tree trunks in groups of two or three, but are sometimes found individually.

Bugs as agricultural pests

Many bugs are considered agricultural pests. In Malaysia, several species are serious pests of rice, including the green planthoppers (*Nephotettix* spp.), brown planthopper (*Nilaparvata lugens*), *Sogatella furcifera*, *Scotinophara coarctata*, *Leptocorisa* spp. and *Nezara viridula*.

Among the five recorded species of *Nephotettix*, four are found in Peninsular Malaysia. *N. impicticeps* is of greatest economic importance because this insect has been shown to be a vector of red disease (tungro) in rice (when it is known as red disease virus). Adults and nymphs of *Nilaparvata lugens* and *Sogatella furcifera* suck the sap of rice, especially the base of rice stems. If the population of *N. lugens* is very high, then the severe attack causes hopperburn, in which the plant becomes yellow and dry.

This group of insects also attacks other economically important crops. For instance, the mosquito bug (*Helopeltis theobromae*) attacks cocoa plants. Both the adults and nymphs cause damage by sucking the plant sap from the young shoots and leaves of the cocoa plants.

The beak of a bug

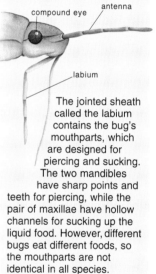

The jointed sheath called the labium contains the bug's mouthparts, which are designed for piercing and sucking. The two mandibles have sharp points and teeth for piercing, while the pair of maxillae have hollow channels for sucking up the liquid food. However, different bugs eat different foods, so the mouthparts are not identical in all species.

Rice pests

Several bugs are known to be major pests of rice. The damage they cause, not only through their feeding habits of sucking the sugary sap from the plant, but also through the devastating diseases they transmit, is enormous. While natural predators of these bugs, such as beetles and parasitic wasps, may help keep numbers down to reasonable levels, farmers may be forced to resort to chemical sprays and other man-made deterrents to keep these fast-breeding pests under control.

The green planthopper (*Nephotettix* sp.)

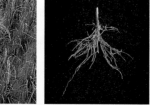

A paddy field attacked by the tungro virus, which damages the roots of rice plants (right). The disease is carried by the green planthopper (*Nephotettix impicticeps*). Hopperburn caused by the brown planthopper (*Nilaparvata lugens*) (inset).

Grasshoppers and crickets

In many countries, the presence of grasshoppers and crickets is feared because they are the cause of massive damage to crops, which can lead to famine. In Malaysia, however, these insects are more noted for their songs and their mimicry than for the damage they inflict.

The bold coloration of *Tauchira polychroa* warns would-be predators of the distastefulness of this grasshopper. The femur of this species has been likened to 'the swell of a dancer's calf'.

Nisitrus sp., a cricket commonly found on Mount Kinabalu. The male cricket sings by rubbing together specialized areas at the base of its wings.

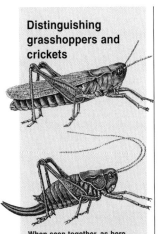

When seen together, as here, the short antennae and barely developed ovipositor of the grasshopper (above) contrast sharply with the long antennae and long ovipositor of the cricket (below).

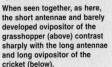

A traditional Malay belief

It is a traditional Malay belief that the *pelesit*, a spirit which takes the form of a grasshopper or cricket, is the pet of another spirit, the *polong*. The pelesit flies to and fro, causing harm only when it enters, tail first, the body of a person chosen by the polong, which then follows.

The pelesit is sometimes kept in a bottle by Malay women who feed it on rice or blood drawn from the tip of the fourth finger. To cure a sick person affected by a pelesit, a *bomoh* or shaman asks the pelesit the name of its mother. The pelesit replies in the patient's voice in a falsetto tone, giving the name of its owner, who is then requested to remove the pelesit.

Leaping insects

Grasshoppers (*belalang*) and crickets (*cengkerik*) belong to the order Orthoptera (from the Greek *orthos* meaning 'straight', and *pteron* for 'wing'). There are more than 10,000 species in the world. Typically, the hind legs are strongly developed and adapted for jumping—the femur is large and muscular. Though most orthopterans can fly, they prefer leaping; hence the alternative name of the order, Saltatoria, from the Latin *saltare* ('to leap'). The forewings (tegmina) are normally long, narrow and leathery. They serve to protect the membranous hind wings and the abdomen, and are often used in the production (by frictional mechanisms) of sounds involved in courtship behaviour. Some species have reduced wings, others none at all.

Differences between grasshoppers and crickets

The order Orthoptera is divided into two broad groups based on the length of the antenna and the ovipositor. Short-horned grasshoppers (typified by the family Acrididae) have short antennae and barely developed ovipositors. The other group has long antennae and long ovipositors which are blade-like or sword-like in bush crickets (family Tettigoniidae) and needle-like in crickets (family Gryllidae) and mole crickets (family Gryllotalpidae). In America, bush crickets are commonly referred to as long-horned grasshoppers; some are called katydids because their songs suggest this combination of syllables.

Most short-horned grasshoppers live among grasses on the ground. All are herbivores (plant feeders) and are active by day (diurnal). Bush crickets are mostly found in the foliage of trees and bushes, but some species live on the ground. They are generally nocturnal, but may be active in the late afternoon. Some are partially predatory, feeding on other insects, for example, the sickle grasshopper (*Conocephalus longipennis*), which is common in paddy fields.

Crickets are usually nocturnal, living in burrows by day. They are omnivorous, eating a wide variety of plant and animal material.

Mole crickets normally live underground. Their food consists of both plant and animal matter. As their name suggests, mole crickets use their highly modified front legs for digging burrows in very much the same way as the mole uses its front limbs. The hind legs of mole crickets are short and thus are not adapted for jumping.

Stridulation and songs

Grasshoppers and crickets, together with cicadas of the order Homoptera, are best known for their melodious songs and sounds. Generally, it is the males who 'sing' to serenade the females and to rival other males of their own kind. Each species has a distinctive song, but the nature of the melody varies according to the effect that is desired. Some females

The life cycle of grasshoppers and crickets

First instar: 8 mm

Grasshoppers and crickets have an incomplete metamorphosis; the egg hatches into a small nymph which resembles a miniature adult, but lacks wings and functional gonads. The nymphs usually have the same diet and habitat as the adults. With each moult (ecdysis) their size increases. Because their wings develop externally, they are called exopterygotes as are other insects with this kind of development.

Female short-horned grasshoppers lay eggs in the ground. The number of eggs laid depends on the species: 70–100 in *Valanga nigricornis* and 12–42 in the oriental migratory locust (*Locusta migratoria manilensis*). They are enclosed in an egg pod. Bush crickets usually lay a single egg, without covering, in the ground, crevices of bark, stems or leaves. Their sword-like ovipositors are used for boring holes in the ground or cutting slits in plants.

wing bud

Third instar: 20 mm

developing wing

Fifth instar: 50 mm

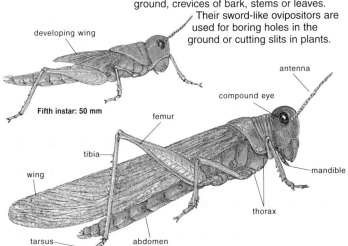

antenna

compound eye

femur

tibia

wing

mandible

thorax

tarsus

abdomen

Early instar nymphs of many grasshopper species display the 'safety-in-numbers' principle.

can sing, but their stridulatory apparatus is poorly developed compared with the males. Both sexes, however, have efficient hearing organs. In the short-horned grasshoppers, the 'ears' are located on the last segment of the thorax or at the base of the abdomen. In bush crickets and crickets, the ears are found on the tibiae of the front legs.

Special structures are developed to produce the sound by stridulation. Males of short-horned grasshoppers stridulate by rubbing a file (a small peg-like projection) on the hind femur over a scraper (a specially hardened vein) on the forewing. Crickets and bush crickets have both the file and scraper on the wings. They are able to produce a shrill, sustained sound by rubbing the forewings together. This sound is amplified by the raised wings acting as resonators. Although some species of crickets produce a deafening noise, the songs of others are inaudible to the human ear.

1. The oriental migratory locust (*Locusta migratoria manilensis*) is the only species of grasshopper in Malaysia that forms large migratory swarms.

2. *Oxya hyla* is a short-horned grasshopper of the rice ecosystem.

3. Feeding on a wide range of plants, *Valanga nigricornis* is a common short-horned grasshopper of cultivated fields and home gardens.

Camouflage and mimicry

Many grasshoppers are masters of the art of disguise and mimicry—resembling something else so their true identity cannot be recognized. The adult of the *Condylodera tricondyloides* grasshopper, living on the jungle floor, closely resembles the predatory tiger beetle *Tricondyla*, even in its mode of running. Very young *Condylodera* grasshoppers are, however, found in flowers together with the beetle *Collyris sarawakensis*, which corresponds in size and coloration to the young grasshopper. Another example of effective camouflage is the bush cricket *Tegra novaehollandiae* which has mottled markings which resemble lichen-covered tree bark.

Some grasshoppers and crickets have evolved a leaf-like appearance. Bush crickets of the genus *Phyllomimus* (right) are superb mimics of leaves, with leaf veins and disease blotches in their repertoire of deception. Just as ingenious and extraordinary is *Chlorotypus gallinaceus* (below), which fully resembles a dried, dead leaf. Populations of some species even comprise

both dead-leaf and green-leaf imitators.

Although their general colouring is mimetic and merges with the surroundings, many grasshoppers have brightly coloured hind wings. When they are disturbed, they combine leaping with flight to unfold the hind wings, which suddenly transforms them into a flash of colour. Some grasshoppers also display warning or aposematic colours which convey the message of distastefulness.

Pests

Only a small number of grasshoppers and crickets damage crop plants in Malaysia. In the rice ecosystem, the most common short-horned grasshopper is *Oxya japonicum*. This is followed by *Atractomorpha crenulata*, which is known as the nose grasshopper because of its strongly elongate, pointed head. Of the long-horned grasshoppers, *Conocephalus longipennis* is the most abundant species. It is possible that flower-eating species, such as crickets of the genus *Nisitrus*, may achieve pest status in the horticulture industry.

The most devastating of orthopterans are the locusts, known since biblical times as one of the greatest menaces to humankind. In migratory swarms, plague locusts, of which there are several species, cause catastrophic damage to crops and bring about famine. Attacks on such a scale are, however, unknown in Malaysia. Although the oriental migratory locust (*Locusta migratoria manilensis*) sometimes causes extensive damage to cultivated crops as well as *Acacia* plants in Sabah, it does not seem to be established in Peninsular Malaysia.

Another potential pest is the black-and-yellow valanga grasshopper (*Valanga nigricornis*) which, in addition to its attacks on plants in suburban gardens, may cause severe defoliation to oil palm and rubber plantations during outbreaks. However, it does not form swarms typical of locusts.

Tin animal money

Traditionally, tin miners employed a *bomoh* or shaman to supervise the smelting procedure to ensure there was harmony between the local spirits and the demons. At a newly opened tin mine, the first ore to be smelted was cast into a pair of 'shell-backed ingots'. When a head and legs were added, the ingots resembled a tortoise. Other animal shapes, which included the grasshopper, were later introduced. These ingots were used in many magical ceremonies and also as a form of currency.

An example of tin money in the shape of a grasshopper.

Cockroaches, houseflies and blowflies

In Malaysia, there are seven species of cockroaches commonly found in homes and buildings: the American, German, brown, Australian, harlequin, brown-banded and lobster cockroaches. The American cockroach is the predominant species in homes, while the German cockroach is a serious pest in hotels and food preparation industries. Houseflies and blowflies are not as numerous in Malaysia as in many other tropical countries, perhaps because there is not a marked dry season.

The harlequin cockroach (*Neostylopyga rhombifolia*) is one of the cockroach species known to be wingless throughout all its life stages.

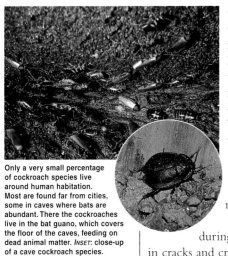

Only a very small percentage of cockroach species live around human habitation. Most are found far from cities, some in caves where bats are abundant. There the cockroaches live in the bat guano, which covers the floor of the caves, feeding on dead animal matter. *INSET*: close-up of a cave cockroach species.

An egg case (ootheca) of an American cockroach (*Periplaneta americana*).

Cockroaches

Cockroaches (*lipas*) are primitive insects belonging to the order Dictyoptera. Fossil evidence suggests that these insects have existed for more than 300 million years. About 4,000 species of cockroaches have been identified throughout the world, but only 1 per cent of them are household pests. The others are found away from human habitation, living in places such as caves and the rich vegetation in rainforests.

Household cockroaches are active during the night, and hide during the day in cracks and crevices in walls, door frames, furniture, cupboards, bathrooms, kitchens, electrical appliances, boxes and drains. They move freely from building to building and from drains and sewers to human dwellings. Cockroaches feed on almost any kind of foodstuff available, although they prefer starchy or sugary types of food. They usually regurgitate some of their partially digested food and defecate while they are feeding. Due to their activity and feeding behaviour, cockroaches are potential mechanical carriers of disease. Various pathogenic organisms have been isolated from the bodies of cockroaches, including bacteria which cause food poisoning, pneumonia and typhoid.

Cockroaches undergo incomplete metamorphosis in their life cycle, which consists of three stages: the egg, nymph and adult. The eggs are laid in a bag-like structure called an ootheca, which is gradually pushed out of the body of the female as it is filled with eggs, and deposited in a crevice. The nymphs emerge from the hatched ootheca, resembling the adults, except that they are smaller in size, wingless and possess undeveloped sex organs. Several moults usually occur before nymphs mature into adults.

Cockroaches are extremely well endowed with sensory receptors, so they are able to move about in the dark. Their long antennae detect humidity, temperature and air pressure. Receptors in their legs pick up ground vibrations, allowing them to make a very speedy escape from any threat of danger.

Some species have wings; others are wingless. However, even the winged species do not all fly.

Species in Malaysia

The American cockroach (*Periplaneta americana*) is found in practically all Malaysian homes. The adults measure 35–40 millimetres in length; females can be distinguished by their stouter abdomen. The length of their wings is also distinguishable from those of males. Wings extend slightly beyond the abdomen in males, but are approximately as long as the abdomen in females. The adult life span ranges from nine months to two years. Many Malaysian Chinese claim that this species has medicinal value for the treatment of asthma in young children.

The German cockroach (*Blattella germanica*) is a small species, about 10–15 millimetres long, with an adult life span of 4–6 months. The males are light yellowish brown, while females are darker. Nymphs are generally black. The German cockroach is the principal problem of the Malaysian pest control industry as it is abundant in urban areas, mostly in hotels and restaurants.

Common household cockroach species

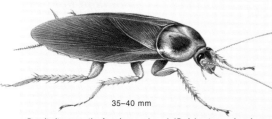

35–40 mm

Despite its name, the American cockroach (*Periplaneta americana*), one of the cosmopolitan species which has spread throughout the world, originated from Africa.

10–15 mm

One of the smallest cockroaches commonly found in Malaysia is the German cockroach (*Blattella germanica*), another of the cosmopolitan species.

30–35 mm

Common in rural areas, the brown cockroach (*Periplaneta brunnea*) displays the cockroach's characteristic flattened, oval-shaped body and hardened wing casing.

The brown cockroach (*Periplaneta brunnea*) and the Australian cockroach (*Periplaneta australasiae*) are commonly found in rural households in Malaysia. Morphologically, both species are quite similar to the American cockroach. The Australian cockroach can be found both indoors and outdoors, as it prefers to feed on vegetation. The male brown cockroach is the only male cockroach which can fly.

The name of the harlequin cockroach (*Neostylopyga rhombifolia*) is probably derived from the intermingled stripes and patches of yellow on its body. Unlike other common Malaysian cockroach species, the harlequin cockroach is wingless, even at the adult stage. When disturbed or squeezed, it produces a sticky, odorous substance.

The brown-banded cockroach (*Supella longipalpa*) is a small species, similar in size to the German cockroach. It is also called the TV cockroach because it can often be found inside old television sets. This species prefers a dry habitat, which explains why it is often found in bedrooms. It is a better flier than the other species.

The lobster cockroach (*Nauphoeta cinerea*) owes its name to the lobster-like design on its back and to its ashy colour. In Malaysia, it is commonly found in sundry shops and food storage sheds. One special characteristic of this species is that it incubates its eggs in its body and then lay its nymphs.

Malaysian flies

Houseflies (*lalat*) (*Musca domestica*) and blowflies (*langau hijau*) are common nuisance insects which are often associated with garbage and other filth. However, they are also found in kitchens where they alight on both raw and cooked food. Like cockroaches, they are potential mechanical carriers of diseases such as typhoid, cholera and dysentery. In addition, several species of blowflies can cause myiasis (infestation of a living body with fly maggots) when they lay their eggs in wounds of humans or animals.

The ability of flies to cling to many types of surface is aided by the pair of 'claws' as well as a fleshy pad on the foot of each of its six legs. The claws help them cling to rough surfaces, while the

Housflies feed on anything they can find, whether rubbish, a bowl of sugar or a freshly prepared, uncovered dinner.

pads are used on smooth surfaces, including ceilings.

Flies are members of the order Diptera, which means 'two-winged insects', with about 80,000 species throughout the world, including Antarctica. However, it is in warm, moist climates that they are most abundant. Unlike cockroaches, they undergo complete metamorphosis. Their life cycle consists of the egg, larval (maggot), pupal and adult stages. Flies usually breed in a wide range of rotting organic matter, provided it is moist but not liquid, including dung, garbage and processing waste.

The housefly is about 6 millimetres long with four distinct dark stripes on the thorax. Its whitish, banana-shaped eggs are laid using a telescopic ovipositor, usually in garbage. The larvae which hatch from the eggs usually undergo three moults, and migrate to drier and cooler places before turning into barrel-shaped pupae, which darken as they become older. The pupal stage lasts 4–5 days before the adult fly emerges. Adult flies prefer fluid and semi-fluid food as the mouthparts or proboscis (a trunk-like organ with a two-lobed tip) are specially designed for sucking and lapping up such food.

The adult male can be distinguished from the female by its eyes, which are larger and are closer together than those of the female.

Houseflies can be found almost anywhere. By day, they are commonly found in kitchens, hawker stalls, shops, restaurants, animal farms, etc. They are usually active during the daytime, with peak activity occurring at noon.

The blowfly is also sometimes known as the latrine fly because it usually lay its eggs on human or animal manure. However, eggs are sometimes laid in wounds of people or animals. The blowfly is less domestic than the housefly, and it seldom settles on people. At 8 millimetres long, the blowfly is slightly larger than the housefly, with an attractively coloured metallic green or blue-green body. Its life cycle is similar to that of the housefly. When the flies are numerous, their humming sound can be audible from some distance away.

Although blowflies are usually associated with dung, this blowfly was photographed on a flower.

There are many more species of non-commensal cockroaches than of household pests. Some species blend with their habitat, such as the one above, which resembles a leaf, while others are colourful, such as the one below.

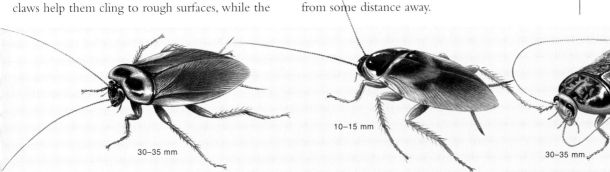

10–15 mm

30–35 mm

30–35 mm

Another rural species is the Australian cockroach (*Periplaneta australasiae*). Not as big a household pest as some species, it prefers to feed on decaying vegetable matter.

Of similar size to the German cockroach, the brown-banded cockroach (*Supella longipalpa*) prefers a dry environment, and is often found in old television sets.

Like other species, the lobster cockroach (*Nauphoeta cinerea*) has mouthparts which face backwards, rather than down or forwards like more advanced insects.

Mosquitoes

Because of its tropical climate, Malaysia is a conducive breeding ground for various mosquito species. Although about 434 species of mosquitoes have been found in the country, only about 10 per cent of them are involved in the transmission of diseases. Important genera include Aedes, Anopheles, Culex *and* Mansonia.

Biological control of the *Aedes* species can be achieved by breeding the voracious larvae of the predator mosquito *Toxorhynchites splendens*.

Classification

Mosquitoes (*nyamuk*) are two-winged insects classified under the order Diptera (family Culicidae). They are important insect pests which feed on humans and animals for blood, and are vectors for many dreaded diseases, such as malaria, filariasis (elephantiasis), dengue, yellow fever and Japanese B-encephalitis.

Metamorphosis

Mosquitoes undergo complete metamorphosis, with a life cycle including egg, larval, pupal and adult stages. The first three stages are usually spent in water, although eggs of several species (for example, *Aedes aegypti* and *Ae. albopictus*) may be laid on moist surfaces. Eggs are ovoid in shape and, depending on the species, are laid singly or bound in the form of ovoid-shaped rafts. A larva emerges from each egg and moults four times before becoming a pupa. The larva is an active form which usually moves by jerking its body.

The larva feeds by rasping at hard materials, scraping algae, ingesting small crustaceans or filtering suspended particles from the water. Generally, it breathes through its respiratory tubes or siphons situated at the posterior of its body, although the *Anopheles* species do not have siphons (they breathe by maintaining their position near the water surface). During respiration, the larva of many species rests at an angle to the water surface. In *Anopheles* species, however, the position is parallel to the water surface.

The pupal stage in mosquitoes, unlike in many other insects, is a mobile stage. However, the pupa does not feed. It is buoyant, due to the presence of float hairs on the first abdominal segment.

Stretches of stagnant water in swamps or flooded plantations are some of the main breeding grounds for mosquitoes.

Fogging with insecticide is one way of controlling the mosquito population.

The blood meal

Once on the host, the mosquito explores the skin with the sensory palps of the labium for 3–10 seconds. When a suitable position is found, the insect presses down with its proboscis, and by a series of minute thrusts the stylets enter the skin and puncture a blood vessel. The labellum remains on the skin's surface.

1. Stylets in blood capillary
2. Skin of host
3. Labium pushed back
4. Pharyngeal pump
5. Salivary gland
6. Stomach
7. Crop
8. Small intestine
9. Rectum
10. Anus

The adult mosquito

The adult emerges from the pupa after 1–3 days (depending on the species). Both the males and females feed on plant juices, particularly tree sap or flower nectar. In order to reproduce, the female requires blood for the development of its eggs. It is generally believed that the female mosquito locates its hosts through smell, detection of carbon dioxide emitted from the host, the body temperature of its host and visual sighting.

Some species can produce eggs without a blood meal (for example, *Toxorhynchites splendens*, commonly known as the predator mosquito). Both sexes in this species feed on tree sap and flower nectar. At the larval stage, they are voracious predators of other mosquito species. Many studies have reported *Toxorhynchites splendens* as a potential biological agent against the *Aedes* mosquito.

Mosquitoes and disease

The female mosquito plays an important role in the transmission of disease from one host to another. When the mosquito feeds on an infected host, the pathogen (disease-causing organism) in the host is transferred into the body of the mosquito through the blood. This pathogen multiplies and undergoes development into an infective stage in the body of the mosquito. When the mosquito feeds on the second host, the infective pathogen is transferred into the blood system of this host.

Mouthparts of the mosquito

The biting mouthparts of the blood-feeding female mosquito consist of six separate elongated parts known as stylets or fascicles, which lie within a flexible, hollow sheath called the labium. The stylets consist of paired mandibles and maxillae which are used to cut into the skin of the host, the tube-like labrum-epipharynx, through which the blood is drawn, and the hypopharynx, down which saliva flows.

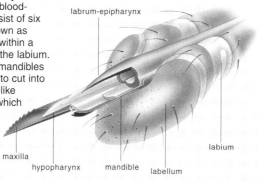

labrum-epipharynx

maxilla

hypopharynx

mandible

labellum

labium

Some vectors of infectious diseases

The *Aedes albopictus* mosquito, one of the two vectors for dengue fever.

Anopheles dirus is one of the many species which carry malaria.

Culex quinquefasciatus, Malaysia's common house mosquito.

Mansonia uniformis is one of the carriers of human filarial parasites.

Aedes mosquitoes

Two species, *Aedes aegypti* (yellow-fever mosquito) and *Ae. albopictus* (Asian tiger mosquito), are vectors for dengue fever in Malaysia, while the former is also a vector for dengue haemorrhagic fever, which can be fatal.

Ae. aegypti breeds mainly in man-made containers with clean water in urban and suburban areas. *Ae. albopictus*, on the other hand, is ubiquitous, and breeds in natural and artificial containers such as tree holes, tin cans, tyres and rock pools. Both species are daytime biters, preferring to bite during the change of light, especially after sunrise and before sunset. Their flight distance is relatively short (approximately 50 metres). *Ae. aegypti* can be differentiated from *Ae. albopictus* by the number of white bands on the thorax. *Ae. aegypti* has three bands, while *Ae. albopictus* has only one band.

Anopheles mosquitoes

There are several species of *Anopheles* involved in the transmission of malaria: *An. maculatus*, *An. dirus*, *An. campestris*, *An. donaldi*, *An. letifer*, *An. balabacensis* and *An. sundaicus*. Their habitat varies according to species, ranging from mountains to coastal areas. They usually bite at night. In Peninsular Malaysia, *An. maculatus* is the principal vector for malaria. At the Malaysia–Thailand border, the key vector is *An. dirus*, while in Sabah and Sarawak it is *An. balabacensis*. When resting, *Anopheles* mosquitoes usually form an angle between their body and the surface which looks like a projecting rocket.

Culex mosquitoes

The main *Culex* species in Malaysia is *Culex quinquefasciatus*. It is the common house mosquito, brown in colour, which usually bites in Malaysian homes at night. In some countries, this species is important as a vector of human lymphatic filariasis (elephantiasis) caused by the parasite *Wuchereria bancrofti*, but in Malaysia the vector is *Anopheles* mosquitoes. *Culex* mosquitoes breed in polluted water, especially in drains, septic tanks, pit latrines and oxidation ponds. Other *Culex* species, such as *C. tritaeniorhyncus*, *C. gelidus* and *C. vishnui*, are vectors for Japanese B-encephalitis. Most *Culex* species are night-time biters.

Mansonia mosquitoes

Several *Mansonia* species, such as *M. uniformis*, *M. indiana*, *M. bonneae*, *M. dives*, *M. annulifera* and *M. annulata*, are vectors for Brugian filariasis, caused by the parasite *Brugia malayi*. Their habitats are mainly irrigation ditches and forest swamps where host plants are present. The immature stages (larvae and pupae) need to attach to the roots of host plants (for example, the water hyacinth) for respiration. *Mansonia* mosquitoes bite mainly at night.

Dengue fever

Although dengue has only become a significant health problem since World War II, the first reported epidemic was over 200 years ago, in 1780.

Dengue is caused by four virus serotypes (called DEN-1, DEN-2, DEN-3 and DEN-4) belonging to the genus *Flavivirus*. The dengue-causing viruses were not isolated until research conducted during World War II. Dengue is transmitted to man by *Aedes aegypti* and *Ae. albopictus* mosquitoes. The incubation period is about 5–8 days. It causes symptoms of aching joints and muscles, fever, a sore throat and an irritating rash. The more severe form, dengue haemorrhagic fever, may be fatal.

A disease of tropical urban areas, in contrast to rural-based malaria, dengue is found in Africa, Asia, Australia and Central America. It is transmitted in a cycle between *Aedes* mosquitoes and humans. It cannot be transmitted from one human to another, though it can be passed from an infected mother to her unborn child. Although infection by one serotype gives future immunity to that serotype, infection by other serotypes is possible. Unlike malaria, dengue patients do not suffer relapses.

The first epidemic of dengue haemorrhagic fever in Southeast Asia occurred in the 1950s. Strict vector control measures prevented more major outbreaks until the 1970s, but with the low-level chemicals used since then, as well as the increasing urbanization and concurrent population increase, the situation has deteriorated and dengue is now the most important mosquito-borne viral disease affecting humans.

A coloured transmission electron micrograph of negatively stained dengue virus particles (red circles).

About 2.5 billion people throughout the world are estimated to live in areas at risk from epidemic transmission. Those most at risk are people living in areas of substandard housing with inadequate waste disposal systems. The growth of air travel has led to cases being reported in an increasing number of countries as people infected by the virus in one country fly on to another country before symptoms are detected.

As yet there is no vaccine available for dengue fever. The best method of eliminating *Aedes* mosquitoes is considered to be larval source reduction, by an integrated approach of environmental sanitation, insecticide spraying and biological control. People living in dengue-prone areas can assist by eliminating any possible breeding sources for *Aedes* mosquitoes in their homes.

REPORTED DENGUE/DENGUE HAEMORRHAGIC CASES BY STATE (1993–1997)				
	1993	1994	1995	1996	1997
Perlis	5	7	10	38	39
Kedah	21	41	40	134	532
Penang	606	423	424	283	840
Perak	434	175	539	683	1,007
Selangor	973	656	1,213	4,153	4,919
Kuala Lumpur	1,706	436	1,506	5,123	6,088
Negeri Sembilan	45	40	235	340	981
Melaka	36	30	43	72	391
Johor	530	525	1,175	1,805	1,793
Pahang	351	236	528	1,006	712
Terengganu	153	46	86	174	560
Kelantan	61	19	94	109	585
Sabah	224	133	226	113	317
Sarawak	470	366	424	222	665
Total	5,615	3,100	6,543	14,255	19,429

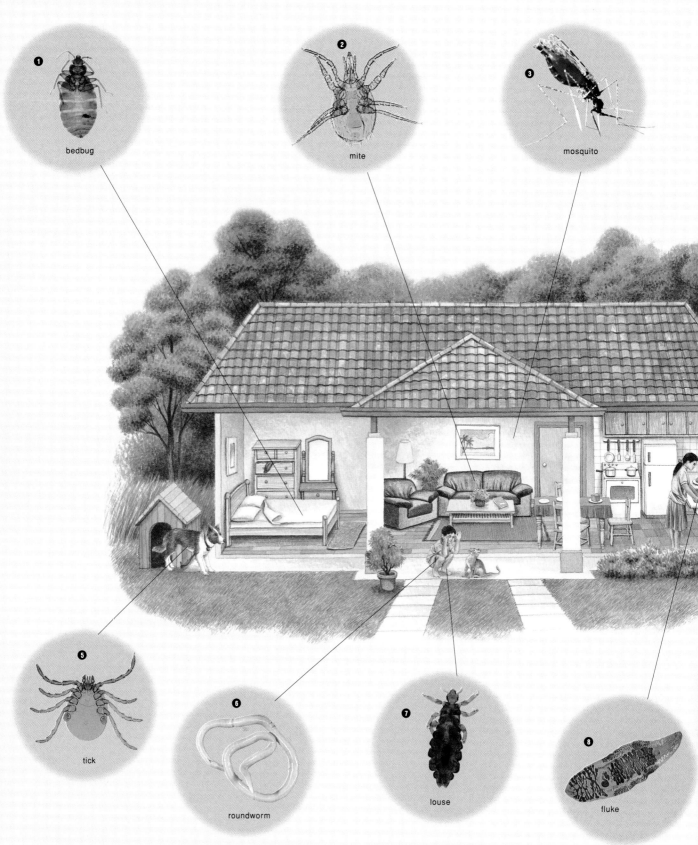

bedbug ❶

mite ❷

mosquito ❸

tick ❺

roundworm ❻

louse ❼

fluke ❽

1. Bedbugs hide in bedding during the day, emerging at night to bite people in bed to obtain a blood meal.

2. Dust mites are often found in carpets and cushions, while other mite species live on birds and humans.

3. Mosquitoes, which require water for breeding, are vectors of the parasites which cause malaria and filariasis.

4. Tapeworms are found in animals such as cattle, but can be transmitted to humans.

5. Ticks are common pests of mammals, birds and reptiles.

6. Many roundworms live in water, soils and plants, but are ingested by humans through poor hygiene or poorly cooked food.

7. Three common types of lice infest humans; they are found on body hair or clothing.

8. Flukes live in snails, fish and crabs, and are transmitted to humans through these foods.

9. Fleas are most commonly found on cats, dogs and rats.

tapeworm

flea

PARASITES

A flea circus, showing trained fleas performing extraordinary acts, viewed with a magnifying glass.

Parasites are typically small organisms that live at least partly, if not entirely, at the expense of another species of living organism, called the host. They may live externally on the body of the host (ectoparasites) or inside the host (endoparasites). Both kinds obtain nutrition, shelter and other requirements from the host, but contribute nothing to the latter's welfare. The adverse effects of parasitism on the host range from almost none to eventual death.

A well-adapted and successful parasite is one which easily infects a host or reinfects another host to carry out its life processes without causing undue or serious harmful effects to its host. The success of any parasite is dependent upon the continued existence of its host.

Some parasites can live in a broad range of hosts, while others can develop in only one particular host species. Indeed, host specificity varies enormously. Human infections acquired from animals are known as zoonoses. Just as the host specificity is highly variable, the parasitic fauna includes a wide range of organisms, from single-celled protozoa to helminths, arthropods and vertebrates. Nearly every phylum has parasitic members. However, the magnitude of the problem and the importance of parasites have not always been given sufficient attention.

Although parasites of humans and animals are often pathogenic (disease-producing), some parasites are of immense potential value in the biological control of pests of agricultural importance.

Bedbugs do not carry diseases, but are merely annoying as they rely on humans for their blood meal. Malaysia has one of the world's two bedbug species—parasites which get their name from their fondness for biting people in bed.

In the past, lice were associated with the transmission of diseases, but this is no longer the case. They are an indicator of poor hygiene and, like bedbugs, are no more than a nuisance.

In contrast, fleas have been the carriers of diseases which have caused many deaths over the centuries. The most serious of these diseases was bubonic plague (the Black Death), which is carried by the oriental rat flea or plague flea (*Xenopsylla cheopis*); an outbreak of this disease occurred in Malaysia as recently as 1928. The other common fleas in Malaysia live on cats and dogs.

Mosquitoes are hosts to both the protozoans, which transmit malaria, and nematode worms, which are the cause of filariasis. Both of these diseases affect wildlife as well as humans. Wildlife are also affected by ticks, which can then be transmitted to both people and domestic animals. Heavy economic losses can be caused to livestock herds in Malaysia affected by tick infestation. However, vaccines are being developed to prevent such losses.

Mites are found on agricultural crops and in houses as well as on wild birds. Those on crops and birds can transmit diseases to humans, who can also be affected by dust mites in houses. Susceptible people can suffer allergic reactions from such mites.

Roundworms, tapeworms and flukes are all found in animals which are eaten by humans. Thus, people are easily infected if such food is not thoroughly cooked. Roundworms also live in soil and on plants, and so the eggs can be easily ingested from dirty hands.

The flea's hind legs enable it to jump 200 times its own body length.

Malaria and filaria parasites

Malaria parasites are protozoans belonging to the family Plasmodiidae, while filaria parasites are nematodes (non-segmented worms) of the superfamily Filarioidea. Both groups contain members which can infect humans, and both are transmitted by mosquitoes. In Malaysia, human filaria infections are caused mainly by Brugia malayi *and* Wuchereria bancrofti, *while malaria infections are caused by* Plasmodium falciparum, P. vivax, P. malariae *and* P. ovale.

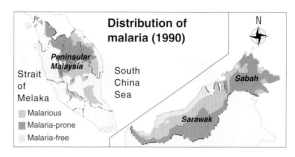

Distribution of malaria (1990)

N

Peninsular Malaysia

Strait of Melaka

South China Sea

Sabah

Sarawak

Malarious
Malaria-prone
Malaria-free

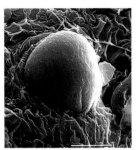

A mature oocyst in the wall of a mosquito mid-gut.

A sporozoite from the salivary gland of a mosquito. *INSET: Anopheles maculatus*, the predominant vector of malaria parasites in Malaysia.

Malaria

Malaria is a serious parasitic disease affecting many tropical and subtropical regions of the world. Human malaria is normally transmitted by *Anopheles* mosquitoes. In Malaysia, malaria is present only in the rural areas, on jungle fringes and in newly opened agricultural land, especially in Sabah.

Malaria parasites

Although human infections are mainly due to four malaria species, there are, worldwide, about 120 species found in primate hosts, rodents, bats, birds and reptiles. Common species infecting the long-tail macaque (*Macaca fascicularis*) in Malaysia are *Plasmodium inui, P. cynomolgi* and *P. knowlesi*. Birds, too, have their own malaria parasites; examples are *P. gallinacium* and *P. formosanum*.

Malaria infection and distribution

A climate with temperatures of 20–32 °C, high humidity and constant rainfall throughout the year provides the best conditions not only for the breeding of *Anopheles* mosquitoes but also the development of plasmodia in them for subsequent transmission of the infection to humans.

In Malaysia, a fairly effective malaria control programme has managed to keep the number of malaria cases to about 40,000–60,000 per year for the last 10 years even though unfavourable factors

like the spread of drug-resistant malaria parasites and vector resistance to insecticides in recent years have worked against control measures.

Mosquito vectors of malaria

Mosquitoes of the genus *Anopheles* can transmit malaria in Malaysia. The predominant vector in Peninsular Malaysia is *An. maculatus*, which breeds in hilly areas in clean streams and seepages open to the sun. In Sabah, the main vector is *An. balabacensis*, which breeds in pools of water under heavy shade in jungle and its fringes. *An. letifer* breeds in coastal plains in stagnant water and *An. campestris* in paddy fields and pits in coconut plantations.

Filaria parasites

Filaria parasites (roundworms) commonly seen in man belong to the genera *Wuchereria, Brugia, Onchocerca, Mansonella, Dipetalonema* and *Loa loa*. Of these, three species—*Wuchereria bancrofti, Brugia malayi* and *B. timori*—cause human lymphatic filariasis (where the parasites are found in the lymphatic system). In Malaysia, human infections are caused mainly by *Brugia malayi* and *Wuchereria bancrofti*. Infections in animals are mostly caused by other species, but *Brugia malayi* does infect leaf monkeys (*Presbytis* spp.).

Mosquito vectors of filariasis

Filaria parasites are mainly transmitted by *Culex, Anopheles, Aedes* and *Mansonia* mosquitoes. In Malaysia, *Wuchereria bancrofti* is transmitted by *Anopheles* mosquitoes. Brugian filariasis, caused by

The transmission of malaria

Female *Anopheles* mosquitoes pick up malaria parasites in the red blood cells from an infected human, but only the sexual stages (gametocytes) are involved in transmission. In the mosquito mid-gut, the male gametocyte produces microgametes which fertilize the female gametocyte, producing a zygote which penetrates the mid-gut wall and lodges itself underneath the membrane covering the wall, where it develops into an oocyst. The mature oocyst produces thousands of sporozoites, which rupture and travel to the salivary glands of the mosquito to await transfer to another human when the infected mosquito next feeds.

The length of time from the infective blood meal to sporozoite migration to the salivary glands is about 10–14 days. Sporozoites

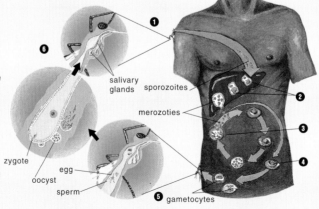

salivary glands sporozoites

merozoites

zygote

oocyst

egg

sperm

gametocytes

The life cycle of plasmodia, the sporozoans which cause malaria

1. A mosquito injects sporozoites.
2. Stages in the liver.
3. Stages in the red blood cells.
4. Certain merozoites develop into gametocytes.
5. Gametocytes are ingested by the mosquito.
6. Sporozoites form within the mosquito.

inoculated into the blood capillaries from a mosquito bite are carried to the liver where they undergo asexual development. This stage usually takes about 7–14 days.

On completion of this stage, thousands of merozoites are released into the surrounding liver tissue. They penetrate the red blood cells, where they undergo development into other asexual stages, finally producing other merozoites which are released on rupture of the infected erythrocyte.

These merozoites can start further rounds of asexual development, which takes about 48 hours in all human malaria parasites except *P. malariae* (72 hours). Some parasites undergo sexual development to produce gametocytes for transmission to another host.

Infection by filaria parasites

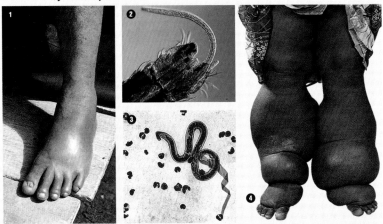

Human lymphatic filariasis is caused by parasitic nematode worms, which are transmitted by various species of mosquitoes. These worms invade the lymphatic system of humans, causing swelling, mainly of the limbs. Chronic infection can result in elephantiasis, with gross swelling of the legs.
1. Early lymphoedema caused by the invasion of the lymphatic system by filaria parasites. 2. The infective stage of filaria worms in the proboscis of a mosquito. 3. A light micrograph of *Brugia malayi* microfilaria in the bloodstream. 4. Elephantiasis due to chronic filariasis.

Malaysia has two forms of brugian filariasis, one mainly a human infection, while the other can be transmitted between humans and animals (mainly monkeys and domestic cats). This is seen in swamp forests and riverine regions, where transmission of the infection from animals to man can occur. The former infection is associated with paddy fields and other agricultural areas.

Animal parasites such as *Dirofilaria repens* and *D. immitis* (dog heart worm) are seen in cats and dogs in some localities, and have occasionally been reported in humans.

1. Swamp forest where numerous water plants can support *Mansonia* mosquitoes breeding.
2. Larvae of *Mansonia* mosquitoes.
3. The *Mansonia* mosquito, a transmitter of the filaria parasite.

Brugia malayi infection, is transmitted by *Mansonia* species breeding on water plants in irrigation ditches, open swamps and swamp forests, and by some *Anopheles* species.

Malaria control in Malaysia

By the beginning of the 20th century, malaria was one of the main obstacles to the economic development of the Malay Peninsula. At that time, malaria was uncontrollable as the cause of the disease was not understood. Land was being cleared at a fast rate for the establishment of towns, mines and especially rubber estates. This wholesale clearing and the lack of sanitation exacerbated the spread of malaria, and led to malaria being the leading cause of death in the Federated Malay States (40–50 per cent in the years 1915–25). In some estates, the number of malaria cases was so high that estates were closed. Not only labourers contracted the disease; European estate managers were also hospitalized.

Moyang kala, a Mah Meri spirit carved on instruction of the bomoh for use in protecting the community from malaria.

The alarming rate of malaria cases in addition to those of beri-beri led to the establishment, in 1901, of the Institute for Medical Research. A British doctor, Malcolm Watson, developed the basic principles of malaria control—drainage, bunding and oiling—which destroyed the mosquito larvae and their breeding grounds. In 1911, the Anti-malaria Control Services and the Malaria Advisory Board were established.

However, control measures were concentrated in the estates, as economically important areas; the other rural areas were neglected. During World War II, the situation deteriorated, and not until the 1960s did a widespread national campaign bring the number of cases down to a manageable level. Constant surveillance has kept the number of cases to about 40,000–60,000 per year.

An integrated approach to malaria control, including early detection and prompt treatment of infected people, vector control and personal protection measures, is used in Malaysia. Surveillance is achieved through detection of infected people in health centres and community surveys. Vector control through anti-larval measures and judicious use of indoor spraying with long-lasting insecticides is employed. In endemic areas, insecticide-impregnated mosquito nets are also used for personal protection and vector control.

Filaria infection

Human and animal filaria infections occur when the infective larvae are deposited on the skin when an infective mosquito obtains its blood meal. These larvae actively enter the puncture wound created by the mosquito's proboscis, and travel to the lymphatics, where they undergo development in the lymphatics and associated lymph nodes, usually of the limbs. When mature, male and female adult worms mate and produce microfilariae (embryos) which enter the blood.

The length of time taken from the infective bite to the presence of microfilariae in the blood is about 8–12 months for *W. bancrofti* and 3–4 months for *B. malayi* infections. When a female mosquito ingests microfilariae from the blood of an infected human, these microfilariae penetrate the mosquito's mid-gut and travel to the thoracic muscles. Here they undergo three developmental stages by the 14th day before migrating to the proboscis to await transmission to another human or animal when the mosquito obtains its next blood meal.

FILARIA PARASITES IN MALAYSIA	
HOST	**SPECIES**
Humans	*Brugia malayi* *Wuchereria bancrofti*
Dogs	*Dirofilaria immitis* *Brugia pahangi*
Cats	*Dirofilaria repens* *Brugia malayi* *Brugia pahangi*
Monkeys	*Brugia malayi* *Brugia pahangi* *Dirofilaria magnilarvatum* *Dipetalonema digitatum*
Chickens	*Cardiofilaria nilesi*

Filariasis control in Malaysia

An effective filariasis control programme since 1961 has reduced the incidence of filariasis in the country. In 1994, only 620 (0.3 per cent) of 224,493 people examined in endemic areas were found to be infected.

Once widely distributed in riverine and swampy areas in Pahang, Perak and Selangor, filariasis cases have been reduced tremendously through a combination of detection and treatment of affected communities with the drug diethylcarbamazine citrate. Filariasis control teams from the Ministry of Health regularly carry out night blood surveys to detect infections and treat affected communities.

In areas with the human form of *Brugia malayi*, the reduction of *Anopheles* vectors resulting from residual insecticide spraying used in malaria control has also benefited filariasis control and many such areas in Kedah and Penang are now filariasis free. Draining of swamps has also effectively reduced *Mansonia* mosquito breeding. Cases of elephantiasis, caused by chronic filariasis, have also been reduced.

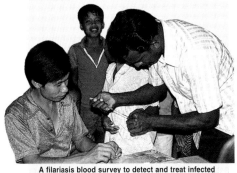

A filariasis blood survey to detect and treat infected communities.

Roundworms, tapeworms and flukes

Roundworms, tapeworms and flukes can all infect humans in Malaysia, as elsewhere, with a number of diseases such as clonorchiasis, schistosomiasis and filariasis. All can be found in animals and can infect people who eat food which is not thoroughly cooked. Roundworms also live in soil, water and plants, and so their eggs can enter humans on unwashed vegetables or even dirty fingers. Transmission to humans can be avoided by close attention to personal hygiene and the consumption of thoroughly cooked food.

Robertsiella kaporensis, a snail intermediate host of the Malaysian schistosome (*Schistosoma malaysiensis*).

This chromolithograph by J. F. Schreiber from *Naturgeschichte* by Dr G. H. von Schubert, published in Germany in the late 19th century, includes drawings of a roundworm (r), tapeworm (w, x) and fluke (u).

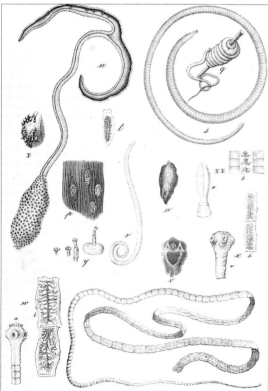

Helminths

'Helminth' is the general term used to describe both parasitic and free-living roundworms as well as tapeworms and flukes. Roundworms, which belong to the phylum Nematoda, are typically elongated and cylindrical in shape. They are covered by a cuticle, below which are the hypodermis and muscle layer surrounding a body cavity in which the digestive, nervous, reproductive and primitive excretory systems are suspended. The sexes are usually separate. Many roundworms are free-living, in water, soil, and the roots and stems of plants. Others are parasitic in humans and animals, causing severe illness and occasionally death. Serious roundworm infestation may cause severe economic loss to the livestock industry.

Tapeworms and flukes belong to the phylum Platyhelminthes and are parasites of animals and humans. As the name implies, tapeworms are normally flat and can be long and tape-like, while flukes are leaf-like in shape.

Roundworms can be large worms, such as *Ascaris lumbricoides*, the common roundworm which infests the intestine of humans and animals; the female can measure about 40 centimetres long by 0.5 centimetre wide. Others are hardly visible to the naked eye, such as *Strongyloides stercoralis* adults which are 2–3 millimetres long and 0.05 millimetre in width.

Medically important because of their parasitic association with humans, roundworms can be found in many different sites, and show extremely diverse modes of transmission. Some, such as *Ascaris lumbricoides* and the whipworm (*Trichuris*

Structure of the large roundworm of humans (*Ascaris lumbricoides*)

mouth
excretory pore
pharynx
intestine
pseudocoel
cuticle
ectoderm
genital pore
intestine
uterus
excretory duct
ovary
uterus
ovary
nerve cord
muscle

A light micrograph of eggs of the nematode worm, *Ascaris lumbricoides*, the common roundworm which infests humans and animals. The mature worm lives in the small intestine, where the female lays large numbers of eggs. These are excreted in faeces and ingested by a new host via contaminated water or food.

A light micrograph of eggs in the ovary of the nematode worm, *Ascaris lumbricoides*.

An egg of *Ascaris lumbricoides*, passed out in human faeces. Eggs are 60–70 micrometres long and take about 2–3 weeks to mature in the soil before they become infective.

trichiura), common human parasites, are transmitted through ingestion of eggs in contaminated food or from dirty fingers. Others, such as hookworms (*Ancylostoma duodenale* and *Necator americanus*), infect humans through penetration of the skin, usually the bare foot, by the infective larval stage of the parasite. These parasitic infections are associated with poor sanitation and personal hygiene. Other roundworms, such as the filaria parasites *Wuchereria bancrofti* and *Brugia malayi*, are transmitted by mosquitoes infected through feeding on other infected humans. The pinworm (*Enterobius vermicularis*) can be transmitted through ingestion of infective eggs on contaminated fingers. Yet other nematode infections are transmitted through the eating of intermediate hosts (such as snails, arthropods or fish which support the larval stages of the parasites and form an essential chain in their life cycle). A good example is *Parastrongylus* (*Angiostrongylus*) *malaysiensis*, which infects humans via ingestion of improperly cooked infected snails or raw vegetables carrying the infective larvae.

Infestations

Worm infestations, such as those due to intestinal roundworms, are usually associated with low socioeconomic situations and poor sanitation. But others are linked with various cultural habits, such as the eating of raw or undercooked meat or fish. Thus, while the common intestinal roundworms are usually seen, especially in children, in rural areas and urban slums, those linked to cultural habits affect people in both urban and rural areas.

A culturally associated helminth infection is clonorchiasis, associated with eating raw fish. This infection is caused by *Clonorchis sinensis*, a fluke transmitted by the eating of some freshwater fish. It is occasionally seen in Malaysia, although it is believed that the infection has been acquired elsewhere. Another fluke, *Paragonimus westermani*, can infect both humans and animals consuming some types of raw or improperly cooked freshwater crabs. The infective stage of the fluke, known as metacercaria, is present in the gills and muscles of crabs. Another important fluke which infects over 200 million people globally is the schistosome, which occurs as various species in different parts of the world but all have snails as the intermediate host. Several schistosomes cause intestinal schistosomiasis, including *Schistosoma intercalatum* and *S. malaysiensis* in Malaysia. The adults lodge in veins, discharging eggs which damage the liver and intestines. The latter parasite infects giant forest rats in Malaysia and is transmitted by various snails found in small jungle streams, including *Robertsiella kaporensis*, a very small freshwater snail.

Tapeworm infections, though common in animals, are uncommon in humans in Malaysia. Tapeworms of rats which can occasionally infect humans are *Hymenolepis nana* and *H. diminuta*. *Taenia saginata*, a tapeworm sometimes seen in humans in Malaysia, is acquired through eating undercooked beef. The larval stage, known as *Cysticercus bovis*, is found in the muscles. If eaten, the adult can develop in the intestine to a length of more than 20 metres.

Life cycle of the human liver fluke (*Clonorchis sinensis*)

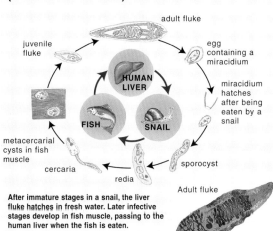

adult fluke

juvenile fluke

egg containing a miracidium

HUMAN LIVER

miracidium hatches after being eaten by a snail

FISH

SNAIL

metacercarial cysts in fish muscle

cercaria

redia

sporocyst

Adult fluke

After immature stages in a snail, the liver fluke hatches in fresh water. Later infective stages develop in fish muscle, passing to the human liver when the fish is eaten.

Structure of the beef tapeworm (*Taenia saginata*)

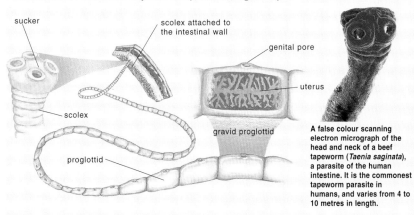

sucker

scolex attached to the intestinal wall

genital pore

uterus

scolex

gravid proglottid

proglottid

A false colour scanning electron micrograph of the head and neck of a beef tapeworm (*Taenia saginata*), a parasite of the human intestine. It is the commonest tapeworm parasite in humans, and varies from 4 to 10 metres in length.

Clinical presentations, treatment and control

In helminth infections, the adult and/or larval stages of the parasite can be present in humans. Adult intestinal worms, such as ascariids and hookworms, are found in the small intestine. The infective larval stage of the dog or cat hookworm (*Ancylostoma ceylanicum*) can penetrate the skin of humans. Other nematode infections, such as the dog roundworm (*Toxocara canis*), can infect humans through soil contaminated with embryonated eggs. If ingested, these eggs hatch out in the intestine and the larvae travel to the liver and other organs, including the brain, where they can cause damage. The rat lungworm (*Parastrongylus malaysiensis*) can also infect man. In human infections, the larval stage of the parasite travels to the brain.

Adult tapeworms in humans do not cause as much damage as the larval stages. The adult *Taenia saginata*, though very large, causes only minimal intestinal discomfort, indigestion, and also interferes with host nutrition. Larval stages, for example those of *Echinococcus granulosus*, can cause severe damage in organs such as the liver and brain.

Adult flukes can also cause severe damage depending on their locality. *Chlonorchis sinensis* can severely affect the bile duct system and liver, and the eggs of schistosomes can cause extensive liver and intestinal damage. Treatment is fairly effective for most intestinal worms. Prevention of helminth infections depends on the parasite involved, but general improvement in socioeconomic conditions, personal hygiene and the provision of proper sanitation can control most intestinal roundworm infections. Other food-borne helminths can be prevented by avoiding undercooked meat, fish and crabs. Control of the intermediate hosts and mass treatment of infected communities are used for the control of schistosomiasis and mosquito-borne nematodes such as filariasis.

A clump of male and female whipworms (*Trichuris trichiura*) taken from the human caecum, a blind-ended pouch at the junction of the large and small intestines. An infestation of these worms is called trichuriasis.

INSET: An egg of *Trichuris trichiura*, found in human faeces. The egg measures about 50 x 20 micrometres and is typically barrel-shaped with a plug at each end. It takes about two weeks to mature in the soil before it becomes infective.

Hookworms

Hookworms, a type of roundworm, cause infections by the penetration of the skin, usually bare feet, by their larvae.

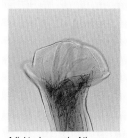

A light micrograph of the copulatory bursa of a male hookworm (*Ancylostoma duodenale*), a human parasite which causes hookworm anaemia. The larvae bore through the skin of the host and migrate to the digestive tract. Eggs of the worm are passed in faeces.

Hookworms attached to the mucosa (internal lining) of the small intestine. Blood loss, leading to anaemia in the host, is due to the sucking of blood by the parasites.

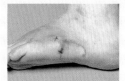

Cutaneous larva migrans due to penetration of the human skin by cat or dog hookworm infective larvae.

Bedbugs, fleas and lice

Bedbugs, fleas and lice are all ectoparasites, living on the surface of their hosts, be they humans or other animals. Malaysia has one of the world's two bedbugs, 800 of the more than 1,500 species of fleas, and the common lice which afflict humans. Advances in medical research have lessened the threat from these parasites since the last outbreak of plague carried by the oriental rat flea in Peninsular Malaysia in 1928.

Though no longer fashionable, flea circuses were displays of trained fleas performing extraordinary acts on microscopic pieces of apparatus which were viewed by the public through a magnifying glass.

The tropical bedbug (*Cimex hemiptera*). When not feeding, the mouthparts are folded beneath the head and thorax. Bedbugs are known to be able to take up to five times their own body weight in blood in one feeding session.

A coloured scanning electron micrograph of a cat flea (*Ctenocephalides felis*) perching on the tip of a hypodermic needle. The insect's flattened body shape can be clearly seen, as can the head, which is immobile and bears simple eyes and biting mouthparts. The well-developed hindlegs are capable of sending the flea distances of up to 20 centimetres.

Bedbugs

The tropical bedbug (*pepijat*) (*Cimex hemiptera*) is often associated with human habitation because it is largely dependent on man for its blood meal. Its name derives from its habit of biting people in bed. The bedbug is ovate in outline, flattened, and measures approximately 5 millimetres in length and 3 millimetres in width. It is reddish brown and emits a distinctive odour. It is a night feeder, hiding during the day in cracks, crevices and mattresses. Usually it can be found in large numbers in its hiding places. Both males and females regularly suck blood from man.

The female lays small, whitish eggs in its hiding place at regular intervals. These eggs hatch into tiny 'mini-adults' (nymphs), which look exactly like their parents except for size. The nymphs undergo five growing stages before becoming mature adults in about five weeks. Each female may lay up to 350–500 eggs in her lifetime. The life span of the adults is about 3–6 months. They are known to withstand starvation for up to a year without feeding.

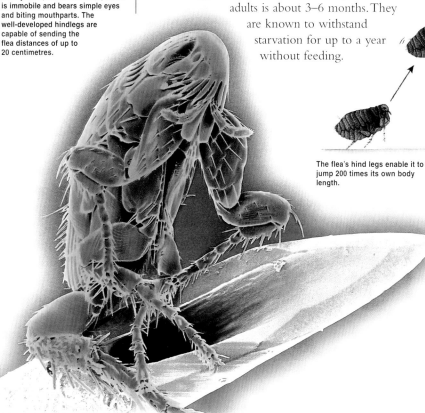

The flea's hind legs enable it to jump 200 times its own body length.

Although bedbugs are extremely annoying because of their bites and also their behaviour of defecating during blood meals, they are not known to be carriers of any disease. Experimentally, they are shown to be able to transmit several diseases, but no such evidence is found in nature.

Though the use of modern synthetic insecticides is most effective in controlling bedbugs, a high degree of personal hygiene and environmental sanitation is the best preventive measure. Bedbug-infested articles, such as clothes and mattresses, must be disinfected.

Fleas as disease vectors

Fleas (*pinjal*) have been responsible for much human suffering throughout history because they are vectors (carriers) of the Black Death (bubonic plague). They belong to the order Siphonaptera, which means 'sucking and wingless insect', because of their blood-sucking habit. Fleas can be easily recognized because they are greatly flattened side-to-side, are usually brown in colour, and readily jump. The female is larger than the male and ingests more blood. The hosts of fleas include a wide range of animals, such as cats, dogs, rats, birds and various types of fowl.

The shiny white eggs of fleas are laid among the hairs of the host or in the nest. Batches of 4–8 eggs are laid at a time, and several hundred eggs may be laid by a single female over a period of 2–4 months. The larvae are legless, with chewing mouthparts. They feed on organic debris and also often on the faeces of the adult fleas. There are three larval stages, and in the last stage the larva builds a cocoon using bits of organic debris and transforms into a pupa. The adult flea then emerges and immediately attaches itself on a new host to suck blood. Depending on the species and the temperature, the flea's entire development may take several weeks to several months. An unfed flea can live up to a year, a fed flea up to five years.

Malaysia's most dangerous flea

A coloured scanning electron micrograph of an oriental rat flea (*Xenopsylla cheopis*) clinging to the fur of its rat host. INSET: Clusters of the bacteria *Yersinia pestis*, cause of bubonic plague, primarily carried by rat fleas.

The most dangerous flea known to have existed in Malaysia is the oriental rat flea or plague flea (*Xenopsylla cheopis*), which is the vector of bubonic plague (or Black Death, because of the black spots seen on the skin). This fatal disease was first seen in Penang in 1896, and last seen in Perak in 1928. The total number of cases within this period was 207. However, in 1933, 765 cases were reported in Singapore, resulting in 712 deaths. The causative agent of bubonic plague is a bacterium (*Yersinia pestis*) which multiplies in the intestine and stomach of the flea and blocks its intestinal tract. As a result, the flea continuously needs to suck blood, since the ingested blood is blocked in its intestine. The blood, which now contains the plague bacterium, is then regurgitated back to the host. In this way, the plague bacterium is transmitted very rapidly. Without medical treatment, death is certain and rapid. Modern treatment with antibiotics, and prevention by immunization, have been successful.

This long life span makes the flea an ideal insect for disease transmission.

The common fleas found in Malaysia are the cat and dog fleas (*Ctenocephalides felis* and *C. canis*). These fleas are not known to transmit diseases to man, but are a source of irritation. The most dangerous flea known to have existed in Malaysia is the oriental rat flea or plague flea (*Xenopsylla cheopis*), which is the vector of bubonic plague.

Flea control is dependent on treating the source, that is, the animal host such as a cat, dog or rat. Dogs and cats can be kept free of fleas by dusting with insecticide powder, or by bathing with an insecticide solution. Rats should be trapped or poisoned, and insecticide powder should be blown into rat burrows. Rat harbourages and hiding places should also be liberally dusted with insecticide. Spraying the house with an appropriate insecticide is also helpful in removing old flea infestation.

Lice on humans

Since ancient times, lice (*kutu*) have been closely associated with man. They are most numerous during times of stress, such as war, famine or disaster, when people are crowded together in unsanitary conditions. Lice inject an irritating saliva into the skin during feeding, causing considerable itching.

Heavy infestation may lead to scratching, bacterial infection and scarred, hardened or pigmented skin—a condition known as pediculosis. In many parts of the world, lice were, in the past, important in disease transmission. However, this is no longer the case. Lice are probably now of minor importance in disease transmission, and should be regarded as nuisances indicative of poor hygiene. Lice are small, flattened and wingless insects with small eyes and mouthparts modified for sucking blood.

There are generally three common types of lice: the head louse, the body louse and the pubic louse. The head louse and the body louse are considered to be two different varieties of the same species, *Pediculus humanus*. The life history of these two varieties is similar. Each female lays over 100 eggs (nits), which are attached to body hair or clothing. These eggs hatch after about nine days to produce nymphs. They take a further nine days to reach maturity after undergoing three moults. Mating takes place within 1–2 days after maturity, and the females live for about a month.

Because of the rapid growth and reproduction, a single fertilized female produces large numbers of eggs in her lifetime. The body louse (*Pediculus humanus*) is known to be associated with the transmission of human diseases, such as epidemic typhus, which is fatal if not treated.

The crab louse (*Phthirus pubis*), the third louse species commonly found on man. Like all louse species, its eggs are glued to the hairs of its host.

The other louse which infests humans is the crab louse (*Phthirus pubis*), which mostly lives in the pubic area but is also often found in other areas such as eyelashes and eyebrows. The female glues her eggs to the hairs, and the entire life cycle is completed within a month. Although it causes irritation, the pubic louse is not known to transmit any disease.

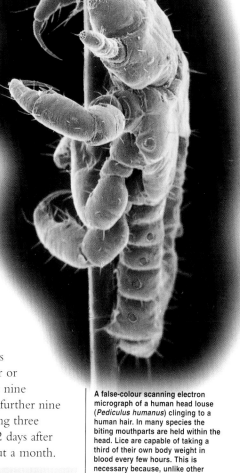

A false-colour scanning electron micrograph of a human head louse (*Pediculus humanus*) clinging to a human hair. In many species the biting mouthparts are held within the head. Lice are capable of taking a third of their own body weight in blood every few hours. This is necessary because, unlike other ectoparasites (such as the flea), a louse can survive for only a few days without regular blood meals. Lice have only simple eyes, or may be completely sightless.

Adults of the body louse (*Pediculus humanus*), which is closely related to the human head louse. A major difference is in habitat: the body louse lives in clothing and crawls on to the host's skin to feed.

Ticks and mites

Ticks and mites have a cosmopolitan distribution and are found in terrestrial, aquatic and marine habitats. The exact number of species in Malaysia is not known, but at least 38 species of ticks have been identified, and there are about 500 recorded species of mites. However, the number of undescribed mite species may be 20 times this number. The majority of ticks and mites have no direct association with man, but some species affect man, his domestic animals and cultivated crops.

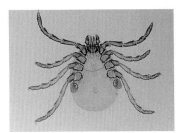

Ticks breathe through a pair of openings located below the last pair of legs. These openings are covered by hard plates called stigmal plates.

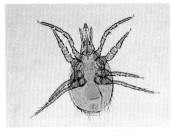

The body of a typical acarine, showing an anterior gnathosoma containing structures for feeding, and posterior idiosoma on which the legs and other organs are attached.

The end of the first leg of a tick, showing a pair of claws and an area of sensory hairs called Haller's organ.

The gnathosoma of a tick showing a hypostome with numerous recurved teeth.

Morphology

Though often confused with insects, ticks and mites form the order Acari, with major characters distinguishing them from insects. Ticks and mites, together with insects, belong to the phylum Arthropoda, which consists of organisms with segmented legs. However, insects belong to the class Insecta, whereas acarines are placed in the class Arachnida, together with spiders. Thus, ticks and mites (which are similar, but ticks are generally larger than mites) resemble spiders more than insects. While insects have three distinct body sections— head, thorax and abdomen— the bodies of acarines are divided into two sections—an anterior gnathosoma, which contains structures for feeding, and a posterior idiosoma, with organs for locomotion, sensation, respiration and copulation. Different families of acarines have different external body structures.

The gnathosoma of all ticks is characterized by a tube-like organ called the hypostome, which is armed with recurved teeth; these teeth enable a tick to attach itself firmly to a host while feeding. Careless removal of ticks from hosts can result in forced separation of the gnathosoma and the idiosoma, with the gnathosoma remaining in the host. The mouthparts of mites are morphologically varied depending on their feeding habits; these may be adapted for piercing, sucking or chewing. Teeth are present on the mouthparts of certain mites and help in grasping.

There are usually four main phases in the life cycle of acarines. Adult females lay eggs which develop into larvae. The larvae further evolve into one or more nymphal stages. Nymphs resemble adults structurally, but are usually smaller and sexually immature. All ticks need to feed on hosts for completion of their life cycle; as such they are major pests of mammals, birds and reptiles. Though all adults and nymphs possess four pairs of legs which resemble the segmented legs of insects, larval

A female tick in the process of laying eggs.

stages of most acarines have only three pairs of legs. The legs of acarines terminate in many forms, such as suckers, claws and elongated hairs.

The sensory organs of ticks and mites are mainly specialized hairs located on many parts of the body which are able to detect chemicals and movements. All ticks have an area on the terminal end of their first pair of legs which contains a group of these sensory hairs; this area is known as 'Haller's organ'. In addition to sensory hairs, some ticks and mites have simple eyes.

The majority of ticks and mites breathe through openings in their bodies called stigmata. Ticks have only one pair of stigmata, located below the last pair of legs and covered by a hard covering called a 'stigmal plate'. One group of mites, with no stigmata, breathe through the entire skin of their bodies. The number and location of stigmata are important clues for the identification of acarines.

There are many modes of reproduction in ticks and mites. There is the usual direct transfer of sperm from the male to the female, and modifications of it. However, certain species of mites possess other sperm transfer organs; or the sperm may be deposited onto a substrate to be retrieved later by the females. While the majority of ticks and mites live in harmony with man, there are a number of species which are of economic and health importance. These ticks and mites affect man and his cultivated crops as well as domestic animals. They cause damage in many ways, including physical damage or irritation during feeding, transmission of pathogens causing diseases, or the production of substances which cause ill health.

Economic and health importance of ticks in Malaysia

All ticks need to feed on a host to complete their development. As common pests of mammals, birds and reptiles, ticks can transmit diseases to man and domestic animals. The species *Ixodes granulatus*, usually found in forests, is an important vector of a number of rickettsial infections affecting man. In the cattle industry, ticks are responsible for causing tremendous economic losses through heavy infestation or through transmission of diseases. One common cattle tick is *Boophilus microplus*. Immature ticks in the forest have also been frequently reported to bite man without necessarily transmitting any pathogens. However, heavy infestation with ticks can also lead to ill health through excessive loss of blood fluids to the feeding ticks. Some ticks secrete a toxin affecting the nerves of the host, resulting in paralysis.

Ticks feed on a snake by inserting their hypostome beneath the scales on the snake's body.

Prolonged attachment of such ticks can be fatal. However, once the tick is removed, the host recovers. Unconfirmed cases of tick paralysis have been recorded in Malaysia.

The ability to multiply in great numbers is common to all ticks, and contributes to their efficiency as pests. Certain species can lay as many as 18,000 eggs. Compared to insects and mites, ticks have exceptional longevity; some ticks have been recorded to survive for several years.

Pesticides are extensively used for the control of ticks. Research workers are conducting studies to produce vaccines to protect cattle against infestation by certain species of ticks. Most of these vaccines are still in the experimental stage, but have proved efficacious. For personal protection when entering an area with known or suspected tick infestation, it is advisable to use a repellent; those containing N, N diethyl-toluamide (DEET) are effective and available commercially.

Economic and health importance of mites in Malaysia

Various species of mites are of public health importance. They are vectors of human diseases, produce materials that are detrimental to human health, or cause irritation by their bite or through their secretions.

The larval stages of some mites of the family Trombiculidae are vectors of a rickettsial disease commonly known as scrub typhus. These larval mites are also called 'chiggers'. Three species of vector chiggers are found in Malaysia: *Leptotrombidum deliense*, *L. fletcheri* and *L. arenicola*. Each is habitat specific. The species *L. deliense* can be found in oil palm and rubber plantations, as well as along forest fringes. *L. fletcheri* is usually found in lalang fields, and *L. arenicola* has been found around sandy coastal vegetation. Scrub typhus is easily treated with antibiotics such as tetracycline and doxycycline. The chiggers can be controlled using pesticides. Repellents such as those against tick bites are also recommended for personal protection if entering an infested area.

House dust mites are a group of mites which form viable populations in house dust; they are found in great numbers in carpets, mattresses and stuffed furniture. Some of these mites produce substances which cause allergic reactions in susceptible humans. Two species found all over the world, *Dermatophagoides pteronyssinus* and *D. farinae*, produce allergens which cause asthma and perennial rhinitis (an inflammation of the nose). The major allergens are located in the faeces of these mites. Medication is recommended for alleviation of the symptoms. Control of the mites is best achieved through integrated measures including vacuuming, use of mite-proof covers for mattresses, and chemical treatment of carpets as well as mattresses.

Scabies, a skin problem caused by mites of the species *Sarcoptes scabiei* which is found in Malaysia, is usually associated with crowded conditions which are conducive for spread of the infestation. The mites burrow into the top layer of the skin. Prolonged infestation causes severe skin irritation, which is especially annoying at night. The problem is self-limiting but in unsanitary conditions it may lead to bad secondary infection by other organisms unless it is treated. In individuals with a compromised immune defence system, such as AIDS patients, a severe form of scabies known as 'Norwegian' or 'crusted' scabies may occur. Treatment can be effected using various miticides which are available commercially. Strict adherence to instructions on application of the miticides is necessary to ensure complete recovery and elimination of the mites. Other forms of *S. scabiei* can cause similar skin problems in domestic and wild animals.

Occupants of houses with roosts of wild birds as well as of poultry farms may occasionally encounter severe itchiness caused by a species of mite called *Ornithonyssus bursa*. This mite usually parasitizes domestic fowls and wild birds. When the nests of wild birds are removed, these mites may enter houses and bite the occupants. Heavy infestation can occur in poultry farms, and may result in people being bitten as well. However, these mites are not known to transmit any pathogens. Treatment of affected premises with residual pesticides is needed to control the problem.

Larval mites (chiggers) of the family Trombiculidae, which are able to transmit the disease scrub typhus to man.

Chiggers can be collected from lalang fields using rectangular black plastic plates.

A false-colour shadow transmission electron micrograph of an individual bacterium of the genus *Rickettsiae*, a group of bacteria which infects ticks and mites, through which they may infect mammals (including man), causing serious diseases, including typhus.

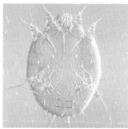
An adult female *Sarcoptes scabiei*, the mite which causes scabies in humans.

Dermatophagoides pteronyssinus, a mite which produces substances which cause allergies. *LEFT*: Various stages. *RIGHT*: An adult female which has been processed so that its outer structures can easily be seen under a microscope.

Small wire traps are commonly used to capture small mammals for studies on acarines found on such mammals.

131

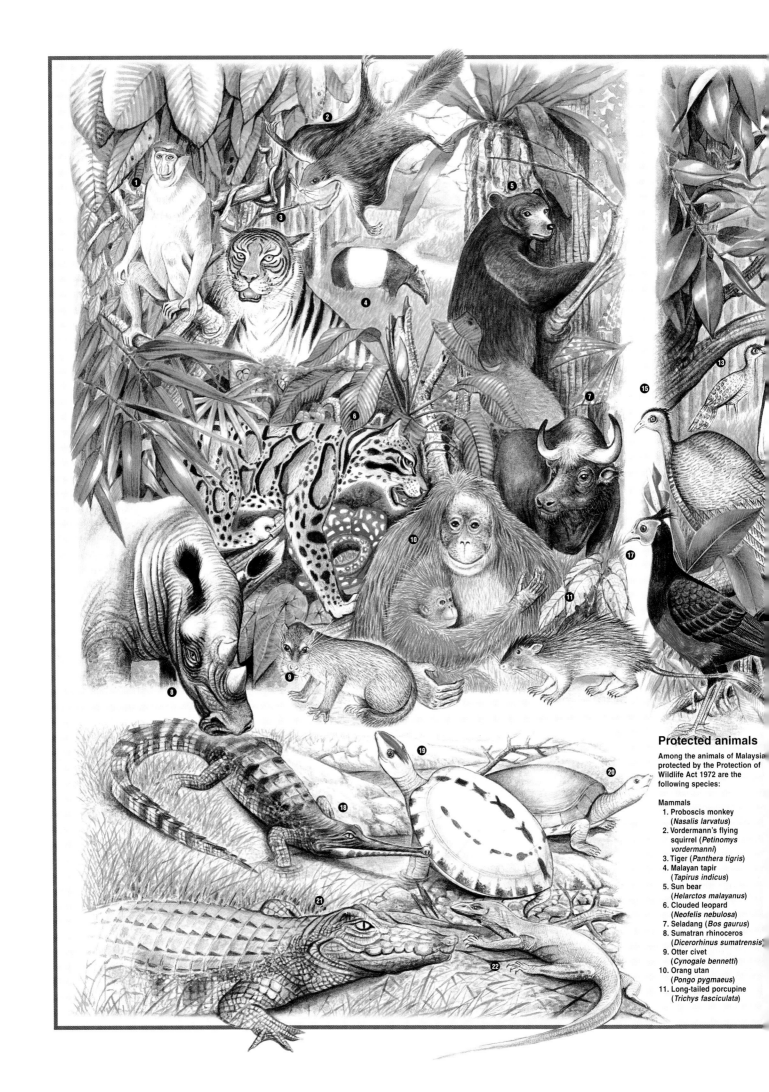

Protected animals

Among the animals of Malaysia protected by the Protection of Wildlife Act 1972 are the following species:

Mammals
1. Proboscis monkey (*Nasalis larvatus*)
2. Vordermann's flying squirrel (*Petinomys vordermanni*)
3. Tiger (*Panthera tigris*)
4. Malayan tapir (*Tapirus indicus*)
5. Sun bear (*Helarctos malayanus*)
6. Clouded leopard (*Neofelis nebulosa*)
7. Seladang (*Bos gaurus*)
8. Sumatran rhinoceros (*Dicerorhinus sumatrensis*)
9. Otter civet (*Cynogale bennetti*)
10. Orang utan (*Pongo pygmaeus*)
11. Long-tailed porcupine (*Trichys fasciculata*)

The crested wood partridge (*Rollulus roulroul*), one of the many game birds which is endangered because of over-hunting.

CONSERVATION

In the last few decades, Malaysia's virgin rainforests have been diminished and replaced by rubber and oil palm plantations as well as settlement schemes. The forests have also been extensively logged. Destruction of the forests and of geological formations, coupled with environmental pollution, are slowly decimating Malaysia's rich natural heritage. In particular, the survival of many of Malaysia's wildlife species is threatened. Legal hunting and illicit poaching of wildlife for sport and commerce, the demands on wildlife for scientific and medical research, the widespread use of pesticides and herbicides, and the needs of zoos as well as the pet trade are all taking further toll on the wildlife of Malaysia. The commercial demands for animal products—skin and fur for clothes, and meat for its purported therapeutic and aphrodisiac values—also threaten the survival and existence of many species.

In Malaysia, the list of declining and endangered species includes a host of mammals, birds, reptiles, amphibians, fishes as well as other little-known wildlife. The last one-horned Javan rhinoceros (*Rhinoceros sondaicus*) is believed to have been shot in 1932. Its relative, the two-horned Sumatran rhinoceros (*Dicerorhinus sumatrensis*), is also on the brink of extinction. The banteng (*Bos javanicus*) is believed to be extinct in Peninsular Malaysia. Hunting has also eliminated one beautiful bird, the Javanese green peafowl (*Pavo muticus*). Other imperilled species can only be assured of a future with conservation programmes.

The question often asked is: Can the declining and endangered species hold their own in the face of the agricultural, industrial and other developments that are taking place at an ever-increasing pace? It is believed that if swift and far-sighted actions are taken in conjunction with careful economic and development planning, Malaysia will be able to perpetuate its splendid natural heritage.

Peninsular Malaysia, Sabah and Sarawak each have separate laws to protect their wildlife. The main laws enacted for this purpose are the Protection of Wildlife Act (Peninsular Malaysia), the Fauna Conservation Ordinance (Sabah) and the Wildlife Protection Ordinance (Sarawak). However, laws alone will not ensure the survival of the species under threat. As long as wildlife exists there will be people determined to flout the law. Furthermore, Malaysia's wildlife species require specific habitats for their existence. Therefore, in addition to laws to protect individual species, the conservation of the whole community, habitat and ecosystem must be emphasized. Once the forest or other natural ecosystem is opened up for development, a great deal of the fauna and flora will perish due to the lack of suitable habitat. Thus, it is vital that more areas be permanently set aside as sanctuaries where Malaysia's natural heritage will be perpetuated.

It is inadequate to introduce laws prohibiting the killing or taking of wildlife. Much more important than laws is educating the general populace about all aspects of Malaysia's natural heritage, including the sustainable use of the natural resources of the country's rich biodiversity. This calls for an effective public extension programme to disseminate information on the value of nature and the importance of respecting it.

A female barking deer (*Muntiacus muntjak*), the smallest of Malaysia's true deer, is included in the list of animals in the Protection of Wildlife Act 1972.

Birds
12. Wrinkled hornbill
    (*Aceros corrugatus*)
13. Malaysian peacock pheasant
    (*Polyplectron malacense*)
14. Lesser adjutant
    (*Leptotilus javanicus*)
15. Great argus pheasant
    (*Argusianus argus*)
16. Milky stork
    (*Mycteria cinerea*)
17. Crested fireback
    (*Lophura ignita*)

Reptiles
18. False gharial
    (*Tomistoma schlegelii*)
19. Painted terrapin
    (*Callagur borneoensis*)
20. River terrapin
    (*Batagur baska*)
21. Estuarine crocodile
    (*Crocodylus porosus*)
22. Clouded monitor lizard
    (*Varanus bengalensis nebulosus*)

In situ and *ex situ* conservation

Malaysia is rated as one of the 12 megadiversity countries in the world because of its rich diversity of flora and fauna. Though the country has only 2 per cent of the total land area of the world, it has 6 per cent of the world's biodiversity. However, rapid development of recent decades has resulted in the reduction of important wildlife habitats, and has become the main cause of the endangered status of some of Malaysia's animals.

Strait of Melaka

0   100   200 km

Taman Negara

Taman Negara, established in 1939 with the name King George V National Park, is the largest protected area in Malaysia, covering an area of 434 300 hectares of rainforest reputed to be 130 million years old. It includes parts of three states—Kelantan, Terengganu and Pahang—and within its boundaries is the highest peak in Peninsular Malaysia, Gunung Tahan (2187 metres). Among the animals which in 1995 attracted 45,000 visitors from 27 countries to Taman Negara are all the large Malaysian mammal species, including the very rare Sumatran rhinoceros, and more than 250 species of birds. It is to Taman Negara that elephants relocated from deforested areas, where they have trampled plantation crops, are brought for

Conservation programmes

The management of endangered wildlife species can be implemented through both *in situ* and *ex situ* conservation projects. *In situ* projects—those designed to protect animals in their natural habitat—include management through surveys, censuses and monitoring of endangered species, such as the Sumatran rhinoceros, seladang, tiger, elephant and migratory birds, in protected areas such as national parks, wildlife reserves and forest reserves. Special management teams have been established by the Department of Wildlife and National Parks to monitor these species in Peninsular Malaysia. In Sabah, wildlife conservation is managed by the Wildlife Department; in Sarawak, the Forestry Department has this role.

Ex situ projects are breeding programmes of animals held in captivity aimed at ensuring the continuation of endangered animal species. The Department of Wildlife and National Parks has at least 15 such projects in Peninsular Malaysia, for animals as diverse as the Sumatran rhinoceros (*badak sumbu*) (*Dicerorhinus sumatrensis*) and pheasants. It also manages Melaka Zoo, to which are taken animals 'rescued' from the wild, such as rogue tigers which have defied attempts at resettlement in the forest. Successful captive breeding programmes are being conducted at this zoo, as they are also at Zoo Negara, which has successfully bred the milky stork (*burung upeh*) (*Mycteria cinerea*), a very rare species of waterbird.

Protected areas

Malaysia's first protected area, the Chior Wildlife Reserve in Perak, was set aside as early as 1903, and there are now 55 such sites throughout the country. The largest protected area is Taman Negara in Peninsular Malaysia, covering an area of 434 300 hectares.

In Sarawak, the largest national park (168 758 hectares) is the Lanjak Entimau Wildlife Sanctuary, bordering the Indonesian state of Kalimantan. A recent survey found nearly 1,000 orang utan (*Pongo pygmaeus*) in the

Tiger mother Melor with her three cubs (April, May and June) which were born in April 1998 in the captive breeding programme at Melaka Zoo run by the Department of Wildlife and National Parks.

reserve, as well as three other endemic Bornean primates: the grey leaf monkey (Hose's langur) (*Presbytis hosei*), the red (maroon) leaf monkey (*P. rubicunda*), and the Bornean gibbon (*Hylobates muelleri*).

The Kinabalu Park in Sabah, covering an area of 75 370 hectares, welcomes more than 100,000 visitors each year, many from overseas. It contains Southeast Asia's highest mountain peak, Mount Kinabalu (4093.372 metres), a favourite of mountain climbers. Although the park does not have many large animals, it has many species of birds, insects and plants endemic to the mountain. In eastern Sabah is a larger park, the Tabin Wildlife Reserve, created for the conservation of the Sumatran rhinoceros.

Some other protected areas are being managed by non-governmental organizations which promote nature conservation and education. Not far from Kuala Lumpur, the Malaysian Nature Society manages the Kuala Selangor Nature Park, which is a bird-watcher's paradise. More than 150 species have been recorded there, including many migratory visitors. Nearby, at Kampung Kuantan is an area of mangrove famous for firefly displays.

Captive breeding programmes

The first captive breeding programme of the Department of Wildlife and National Parks was started in 1968 at Bota Kanan, Perak, to conserve the river terrapin (*tuntung sungai*) (*Batagur baska*). More than 35,000 terrapins have been bred and released into the Perak River as a result of this programme, which has been extended to other locations in Kedah and Terengganu. Another programme has

(Map labels:)
Perlis
River terrapin conservation project — Bukit Pinang
Pulau Pinang
Bird conservation project
Kedah
Kelantan
Perak
Bukit Pakoh
Chior Wildlife Reserve (4330 ha) — Kuala Gula — Gua Musang
Cameron Highlands Wildlife Reserve (64 953 ha)
River terrapin conservation project
Bota Kanan
Terengga
Sungkai Wildlife Reserve (2428 ha) Deer and pheasant captive breeding projects
Selangor
Pahang
Sungai Dusun Wildlife Reserve (1330 ha) Rhino captive breeding project
Negeri Sembilan
Melaka
Kuala Selangor Nature Park (240 ha)
J

Protected areas and conservation programmes

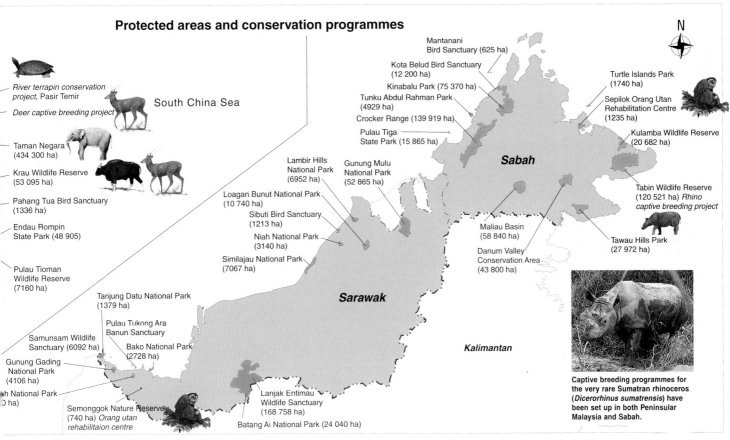

N

River terrapin conservation project, Pasir Temir

Deer captive breeding project

South China Sea

Taman Negara
(434 300 ha)

Krau Wildlife Reserve
(53 095 ha)

Pahang Tua Bird Sanctuary
(1336 ha)

Endau Rompin
State Park (48 905)

Pulau Tioman
Wildlife Reserve
(7160 ha)

Tanjung Datu National Park
(1379 ha)

Pulau Tukong Ara
Banun Sanctuary

Samunsam Wildlife
Sanctuary (6092 ha)

Bako National Park
(2728 ha)

Gunung Gading
National Park
(4106 ha)

h National Park
0 ha)

Semonggok Nature Reserve
(740 ha) _Orang utan
rehabilitaion centre_

Lambir Hills
National Park
(6952 ha)

Loagan Bunut National Park
(10 740 ha)

Sibuti Bird Sanctuary
(1213 ha)

Niah National Park
(3140 ha)

Similajau National Park
(7067 ha)

Gunung Mulu
National Park
(52 865 ha)

Sarawak

Lanjak Entimau
Wildlife Sanctuary
(168 758 ha)

Batang Ai National Park (24 040 ha)

Mantanani
Bird Sanctuary (625 ha)

Kota Belud Bird Sanctuary
(12 200 ha)

Kinabalu Park (75 370 ha)

Tunku Abdul Rahman Park
(4929 ha)

Crocker Range (139 919 ha)

Pulau Tiga
State Park (15 865 ha)

Sabah

Maliau Basin
(58 840 ha)

Danum Valley
Conservation Area
(43 800 ha)

Kalimantan

Turtle Islands Park
(1740 ha)

Sepilok Orang Utan
Rehabilitation Centre
(1235 ha)

Kulamba Wildlife Reserve
(20 682 ha)

Tabin Wildlife Reserve
(120 521 ha) _Rhino
captive breeding project_

Tawau Hills Park
(27 972 ha)

Captive breeding programmes for the very rare Sumatran rhinoceros (*Dicerorhinus sumatrensis*) have been set up in both Peninsular Malaysia and Sabah.

---

since been established for the more colourful painted terrapin (*tuntung laut*) (*Callagur borneoensis*), whose females have the unusual habit of swimming along the river to the beach to lay their eggs.

There are two breeding centres for the Sumatran rhinoceros, one in Peninsular Malaysia at the Sungai Dusun Wildlife Reserve in Selangor, established in 1991, with five rhinos, and another at the Tabin Wildlife Reserve in Sabah, established in 1992, with four animals. With only 77–130 rhinos in Peninsular Malaysia and approximately 40 in Sabah, the future of this species is not assured.

More than 30 seladang (*Bos gaurus*) have been born since 1985 at the Krau Wildlife Reserve breeding centre in Pahang, set up to redress the drastic reduction in numbers of this animal caused by loss of habitat and also poaching. Three centres have been established for breeding sambar deer (*rusa*) (*Cervus unicolor*)—at the Sungkai Wildlife Reserve in Perak, the Krau Wildlife Reserve in Pahang and at Gua Musang in Kelantan. The sambar deer is the largest of Malaysia's deer species, and so was popular with hunters both for its meat and as a trophy. There are now about 120 animals in these three centres, but breeding them is difficult as the fawns are susceptible to diseases.

Orang utan (*Pongo pygmaeus*) rescued from deforested areas are given shelter at Sepilok in Sabah and Semonggok in Sarawak

At Sepilok on Sabah's east coast and at Semonggok in Sarawak are rehabilitation centres where orang utan rescued from deforested areas and those which have been illegally kept as pets are reintroduced to their natural habitat. Once they are able to survive on their own, the orang utan are released into the rainforest. These centres also attract tourists from all over the world as they can see the orang utan in a setting close to its natural habitat.

River terrapin conservation

In 1967, it was realized that the river terrapin (*tuntung sungai*) (*Batagur baska*) was in danger of extinction because almost all the eggs were being collected and sold for consumption. The law requiring that egg collectors buried one-third of the eggs collected was not being strictly enforced as it had been before World War II. During the war, the terrapins were killed for food, seriously depleting their numbers.

A conservation programme was started with the establishment of a hatchery at Bota Kanan, Perak. With experimentation, an efficient method of hatching the eggs was devised. The hatchlings are raised in captivity until they are one year old and then released into the Perak River; about 2,000 are released each year.

Through this programme, biological knowledge of the river terrapin has been increased by studying specimens raised in captivity for this purpose.

River terrapin hatchery

1. A river terrapin in the hatchery.
2. Eggs of the river terrapin.
3. Nursery area of the river terrapin hatchery where the eggs are buried.
4. Releasing year-old terrapin hatchlings into the Perak River.

Glossary

A

Amphibian: Cold-blooded vertebrate typically living on land but breeding in water.
Annelid: Elongate, worm-like animal with body divided into equal rings or segments, such as earthworms and leeches.
Aposematic: Bright, conspicuous markings which predators recognize and learn to avoid.
Aquatic: Living in water.
Arachnid: Chelicerate arthropod, characterized by simple eyes and four pairs of legs such as spiders.
Arboreal: Living in trees.
Arthropod: Animal with jointed legs.

B

Biodiversity: Existence of a wide variety of plants and animals in their natural environments.

C

Camouflage: Resemblance to their surroundings which allows animals to escape the notice of their predators.
Cannabalistic: Eating the flesh of one's own kind
Canopy: Highest level of branches and foliage of trees in forest.
Carnivore: Meat-eating animal.
Chelicerate: Arthropod with two main body sections and a pair of chelicerae in front of the mouth, such as spiders and scorpions.
CITES: Convention on International Trade in Endangered Species.
Commensal: Living in close association with people.
Crepuscular: Active at dusk.
Crustacean: Arthropod with hard carapace, such as crabs.
Cryptic: Tending to conceal by disguising the shape.

D

Decapod: Crustacean with five pairs of walking limbs, such as crabs and prawns.

Dimorphism: Two distinct types of animal in one species.
Diurnal: Active during the day.

E

Echolocation: Determining the position of an object by measuring the time taken for an echo to return from it. Used by bats and cave-dwelling birds.
Ecosystem: System involving interactions between a community and its environment.
Endemic: Species restricted to a certain region, where it probably originated.
Evolution: Gradual change in the population of animals or plants over successive generations.
Extant: Still in existence.

F

Fang: Long, pointed tooth of a poisonous snake through which venom is ejected.

G

Gastropod: Mollusc with a flattened muscular foot for locomotion and a head with stalked eyes, such as snails.
Guano: Dried excrement of cave-dwelling animals such as bats.

H

Herbivore: Vegetation-eating animal.
Hermaphrodite: Organism with both male and female functioning reproductive organs.

I

Insect: Arthropod with body divided into head, thorax and abdomen, three pairs of legs and (in most species) two pairs of wings.
Insectivore: Insect-eating animal.
Instar: Developmental stage of an insect between any two moults.
Invertebrate: Animal without a backbone.

L

Littoral: Inhabiting the shore of a sea or lake.

M

Mammal: Warm-blooded vertebrate which suckles its young.
Metamorphosis: Transformation of a larva into an adult.
Migrant: Animal that moves from one region or country to another.
Mimicry: Resemblance shown by one animal to another, protecting it against predators.
Mollusc: Invertebrate with soft, unsegmented body and often a shell, secreted by a fold of skin (mantle), such as gastropods and bivalves.
Morphology: Form and structure.
Moult: Shedding of feathers, hair, skin or cuticle by birds, mammals, reptiles and arthropods.
Myriapods: Millipedes and centipedes.

N

Nocturnal: Active at night.
Nymph: Immature stage of an insect which lacks a pupal stage.

O

Omnivore: Animal which eats any type of food.
Ovipositor: Egg-laying organ of most female insects.

P

Parasite: Small oraganism that lives at least partly at the expense of another species.
Parthenogenesis: Development of an egg without fertilization, as in aphids and some lizards.
Pheromone: Chemical substance secreted by certain animals, such as insects, which affects the behaviour or physiology of animals of the same species.

Phytophagous: Feeding on plants (usually insects).
Plankton: Free-floating, mostly microscopic, aquatic animals.
Polymorphism: Occurrence of more than one form of individual in a single species.
Predator: Carnivorous animal.
Prehensile: Adapted for grasping, especially for wrapping around a support.
Prey: Animal hunted by another for food.
Pupa: Developmental stage of an insect between larval and adult stages which does not eat or move, and is often wrapped in a cocoon.

R

Reptile: Cold-blooded vertebrate with horny covering which lays eggs.
Riparian: Inhabiting or situated on the bank of a river.
Ruminant: Animal which chews its cud and has a stomach with four compartments, including deer, cattle, sheep and goats.

S

Scavenger: Animal that feeds on decaying organic matter.
Stridulate: Produce sounds by rubbing one part of the body against another, as done by crickets and cicadas.
Sublittoral: Living close to the seashore.

T

Terrestrial: Living on the ground.
Tetrapod: Vertebrate with four limbs.

U

Ungulate: Mammal with hooves.

V

Vertebrate: Animal with bony skeleton and well-developed brain.

Bibliography

Barlow, H. S. (1982), *An Introduction to the Moths of South East Asia*, Kuala Lumpur: Malayan Nature Society.

Bennett, E. L. and Gombek, F. (1993), *Proboscis Monkey of Borneo*, Kota Kinabalu: Natural History Publications.

Berry, P.Y. (1975), *The Amphibian Fauna of Peninsular Malaysia*, Kuala Lumpur: Tropical Press.

Chey Vun Khen (1997), *Forest Insect Pests in Sabah*, Sandakan: Sabah Forestry Department.

Chung, A.Y. C. (1995), *Common Lowland Rainforest Ants of Sabah*, Sandakan: Sabah Forestry Department.

Corbet, A. S. and Pendlebury, H. M. (1992), *Butterflies of the Malay Peninsula*, rev. edn, Kuala Lumpur: Malayan Nature Society.

Cranbrook, Earl of (1991), *Mammals of South-East Asia*, 2nd edn, Singapore: Oxford University Press; first published in 1987 as *Riches of the Wild: Land Mammals in South-East Asia*.

Cranbrook, Earl of (ed.) (1988), *Key Environments: Malaysia*, Oxford: Pergamon Press.

Fetherstonhaugh, A. H. (1940), 'Some Notes on Malayan Bears', *Malayan Nature Journal*, 1: 15–18.

Harrison, John (1966), *An Introduction to the Mammals of Singapore and Malaya*, Singapore: Malayan Nature Society.

Harrisson, Barbara (1988), *Orang-utan*, Singapore: Oxford University Press.

Holloway, J. D. (1976), *Moths of Borneo with Special Reference to Mount Kinabalu*, Kuala Lumpur: Malayan Nature Society.

Hornaday, William T. (1993), *The Experiences of a Hunter and Naturalist in the Malay Peninsula and Borneo*, Kuala Lumpur: Oxford University Press; orignally published as *Two Years in the Jungle*, London: Kegan Paul, 1885.

Inger, R. F. and Chin, P. K. (1952), *The Freshwater Fishes of North Borneo*, Kota Kinabalu: Natural History Publications.

Inger, R. F. and Stuebing, R. B. (1989), *Frogs of Sabah*, Kota Kinabalu: Sabah Parks.

_____ (1997), *A Field Guide to the Frogs of Borneo*, Kota Kinabalu: Natural History Publications.

Inger, R. F. and Tan Fui Lan (1996), *The Natural History of Amphibians and Reptiles in Sabah*, Kota Kinabalu: Natural History Publications.

International Conference on National Parks and Protected Areas, Kuala Lumpur, 13–15 November 1989, Kuala Lumpur: Department of Wildlife and National Parks.

Kaplan, Gisela and Rogers, Lesley (1994), *Orang-utans in Borneo*, Armidale: University of New England Press.

Lim Boo Liat (1981), *Orang Asli Animal Tales*, Singapore: Eastern Universities Press.

_____ (1991), *Poisonous Snakes of Peninsular Malaysia*, 3rd edn, Kuala Lumpur: Malayan Nature Society.

Madoc, G. C. (1976), *An Introduction to Malayan Birds*, 2nd impression, Kuala Lumpur: Malayan Nature Society; first published 1956.

Mak Joon Wah and Yong Hoi-Sen (eds.) (1986), *Control of Brugian Filariasis*, Kuala Lumpur: World Health Organization/Institute for Medical Research.

Malayan Naturalist, various issues

Malayan Nature Journal, various issues

Maxwell, George (1907), *In Malay Forests*, London; reprinted Singapore: Eastern Universities Press, 1960.

Medway, Lord (1978), *The Wild Mammals of Malaya (Peninsular Malaysia) and Singapore*, 2nd edn, Kuala Lumpur: Oxford University Press.

Medway, Lord and Wells, D. R. (1976), *The Birds of the Malay Peninsula*, Vol. 5, Kuala Lumpur:

Penerbit Universiti Malaya.

Mohammad Mohsin, A. K. and Mohd Azmi Ambak (1983), *Freshwater Fishes of Peninsular Malaysia*, Serdang: Universiti Pertanian Malaysia Press.

Mohd Khan bin Momin Khan (1992), *Mamalia Semenanjung Malaysia*, Kuala Lumpur: Department of Wildlife and National Parks.

Nature Malaysiana, various issues

Ng, P. K. L. (1988), *The Freshwater Crabs of Peninsular Malaysia and Singapore*, Singapore: National University of Singapore.

Ooi, P. A. C. (1988), *Insects in Malaysian Agriculture*, Kuala Lumpur: Tropical Press.

Payne, Junaidi and Francis, Charles M. (1985), *A Field Guide to the Mammals of Borneo*, Kota Kinabalu: Sabah Society with WWF Malaysia.

Raffles Bulletin of Zoology, various issues

Reid, J. A. (1968), *Anopheline Mosquitoes of Malaya and Borneo*, Kuala Lumpur: Government of Malaya.

Robinson, G. S.; Tuck, K. R. and Shaffere, M. (1994), *A Field Guide to the Smaller Moths of Southeast Asia*, Kuala Lumpur: Malayan Nature Society.

Schubert, G. H. von. (c. 1886), *Naturgeschichte*, Esslingen and Munich.

Skeat, Walter W. (1900), *Malay Magic*, London: Macmillan; reprinted Singapore: Oxford University Press, 1984.

_____ (1901), *Fables and Folk-tales from an Eastern Forest*, Cambridge: Cambridge University Press.

Smythies, B. E. (1981), *The Birds of Borneo*, 3rd edn, Kota Kinabalu: Sabah Society.

Thapa. R. S. (1982), *Termites of Sabah*, Sandakan: Sabah Forest Department.

Tho, Y. P. (1992), *Termites of Peninsular Malaysia*, Kuala Lumpur: Forest Research Institute Malaysia (FRIM).

Tropical Technical Group (1986),

'Snails of Medical Importance in Southeast Asia', *Southeast Asian Journal of Tropical Medicine and Public Health*, 17: 282–321.

Tung Weng-Yew, Vincent (1983), *Common Malaysian Beetles*, Kuala Lumpur: Longman.

Tweedie, M. W. F. (1970), *Common Birds of the Malay Peninsula*, 2nd edn, Kuala Lumpur: Longman; first published in 1960 as *Common Malayan Birds*.

_____ (1978), *Mammals of Malaysia*, Kuala Lumpur: Longman.

_____ (1983), *The Snakes of Malaya*, 3rd edn, Singapore: National Printers.

Tweedie, M. W. F. and Harrison, J. L. (1970), *Malayan Animal Life*, 3rd edn, Kuala Lumpur: Longman; first published 1954.

Vreden, G. van and Abdul Latif Ahmadzabidi (1986), *Pests of Rice and Their Natural Enemies in Peninsular Malaysia*, Wageningen: Pudoc.

Wallace, Alfred Russel (1869), *The Malay Archipelago: The Land of the Orang-Utan, and the Bird of Paradise*, London: Macmillan; reprinted Singapore: Oxford University Press, 1989.

Wharton, R. H. (1978), *The Biology of Mansonia Mosquitoes in Relation to the Transmission of Filariasis in Malaya*, Kuala Lumpur: Institute for Medical Research.

Whitehead, J. (1893), *Exploration of Mount Kina Balu, North Borneo*, London: Gurney and Jackson.

Whittow, G. Causey (1983), *Malaysian Wildlife: A Personal Perspective*, Petaling Jaya: Eastern Universities Press.

Yap, S. K. and Lee, S. W. (eds.) (1992), *In Harmony with Nature: Proceedings of the International Conference on Conservation of Tropical Biodiversity, 12–16 June 1990*, Kuala Lumpur: Malayan Nature Society.

Yong Hoi-Sen (1983), *Malaysian Butterflies: An Introduction*, Kuala Lumpur: Tropical Press.

Index

Picture Credits

Abdul Hamid Mohd Noor, p. 56, *kris pekaka*. **Abdul Wahid bin Bulin**, pp. 36–7, flying squirrel; p. 96, scorpion; p. 97, centipede, millipede. **Ambu, Stephen**, p. 85, snails of medical importance. **Anuar bin Abdul Rahim**, p. 1, hill myna; p. 5, estuarine crocodile; p. 36, tree; p. 41, owl; p. 113, dung beetle; p. 136, mouse deer; p. 137, squirrel; p. 138, Pacific swallows; p. 144, slow loris. **Antiques of the Orient**, p. 9, game hunters; p. 10, pedlar (from Hugo V. Pedersen, *Door Den Oost-Indischen Archipel*, 1902); p. 98, leeches and earthworms; p. 126, roundworms, tapeworms and flukes. **Auscape International**: Jean-Paul Ferrero, p. 48, white-bellied swiftlets, p. 49, bird's nest collecting, black-nest swiftlet; Joe McDonald, p. 71, Wagler's pit viper; Frank Woerle, pp. 62–3, crocodile. **Chai Kah Yune**, pp. 6–7, animals; p. 13, flying lemur; p. 14, deer, hippo; p. 15, pangolins, elephants; p. 19, monkeys and gibbons; p. 25, bear heads; p. 26, otter, civet and mongoose; p. 27, mongoose and cobra; p. 29, fighting pigs, *tuntun*; p. 31, mouse deer trap, deer and mouse deer; p. 32, pangolins; p. 33, porcupines, porcupine and leopard; p. 35, bat heads and tails; p. 36, squirrel's nest; p. 38, squirrel and treeshrew heads; pp. 38–9, rat, shrew and mouse; p. 45, nightjar spirit; p. 49, bird's nest soup; p. 50, quails; p. 51, metal pillow end; p. 52, pigeon eye design; p. 56, kingfisher in nest; p. 62, crocodile and false gharial heads, crocodile breeding; p. 63, crocodile mask; p. 67, *cicak* motifs; p. 68, flying gecko; pp. 68–9, flying lizard; p. 69, flying snake; p. 70, cobra and pit viper heads; p. 81, walking catfish; p. 83, puffer fish spirit; p. 87, crab, prawn and snail spirits, snail motifs; p. 93, horseshoe crab mask; p. 95, making spider web, spinneret; p. 98, leech basketware pattern; p. 99, sparrow and earthworm; p. 101, bee, fly, mosquito; p. 106, bee spirit; p. 109, termite mound spirit; p 112, beetle and weevil heads, beetle flight; p. 114, cicada development; p. 115, bug's beak; p. 123, p. 128, flea circus, jumping fleas; p. 143, deer head. **Chan Bing Fai**, p. 102, Rajah Brooke's birdwings. **Chan Chew Lun**, p. 100, stick insect,

leaf insect; p. 110, *Pharnacia serratipes*, *Lonchodes hosei*; p. 111, *Heteropteryx dilatata*. **Chang, Tommy**, p. 10, water buffalo; p. 11, Bajau horsemen; p. 50, crested fireback pheasant; p. 60, crocodile; p. 115, man-faced bug, stink bug nymph. **Chey Vun Khen**, p. 102, great mormon, common tiger, blue pansy; p. 103, *Troides amphrysus*, *Pingasa ruginaria*, tree nymph caterpillar, Malay lacewing caterpillar, lymantriid caterpillar; p. 109, *Microcerotermes* termite nest. **Chia, D.**, p. 89, coconut crab, coconut crab abdomen. **Chin Paik Chong**, p. 78, common kingfisher. **China Press**, p. 73, reticulated python. **Chu Min Foo**, pp. 64–5, turtles and tortoises; p. 66, lizard with cracked backbone, gecko head, gecko foot; p. 67, monitor lizard head; p. 68, underside of flying gecko, flying lizard skeleton; pp. 82–3, aquarium fishes; p. 88, hermit crab, fiddler crab signalling; p. 92, dorsal and ventral views of horseshoe crab; p. 101, beehive cross section. **Chung, Arthur Y. C.**, p. 101, ants; p. 108, forest ants; brown crazy ants milking membracid, weaver ant nest, weaver ants making nest; p. 109, *Crematogaster inflata*, termite queen, termite damage to log, *Nasutitermes* termites. **Compost, Alain**, p. 16, male orang utan; p. 18, western tarsier; p. 24, sun bear attacking tree trunk; p. 34, bat flight series. **David Bowden Photographic Library**, p. 30, deer park; p. 80, fishery research centre. **Diesel, R.**, p. 86, punice crab. **Ding, H. O.**, p. 93, holiday-makers. **Diong, C. H.**, p. 26, otter series; p. 66, *Cyrtodactylus malayanus*, tokay gecko, oriental garden lizards (male, female, juveniles), gecko underside, *Gonocephalus chamaeleontinus*, *Phoxophrys nigrilabris*; p. 67, common skink, *Gonocephalus abbotti*, *Gonocephalus grandis*, *Gonocephalus robinsonii*, clouded monitor lizard; p. 68, flying gecko. **Falconer, John,** p. 54, Sea Dayaks by Charles Hose. **Farid Yunus**, p. 42, Langkawi eagle. **Gullick, John**, p. 28, fighting scladang. **Harrisson, Barbara,** p. 17, Barbara Harrisson and orang utan. **Ho Tze Ming**, p. 122, mite, tick; p. 130, tick, typical acarine, first leg of tick,

gnathosoma of tick, tick laying eggs; p. 131, ticks feeding on snake, capturing acarines, chiggers, collecting chiggers, *Sarcoptes scarbiei*, *Dermatophagoides pteronyssinus*. **Hon Photo**, p. 115, cotton stainers. **Hogarth, Ann Novello**, p. 107, bee, wasp and hornet stings; p. 109, termite mound. **Hughes, A. M.**, Prevost's squirrels, p. 15. **International Rice Research Institute**, p. 115, rice pests. **Isao Enomoto**, p. 16, juvenile and female orang utan. **Jacobs, Joseph**, p. 3, sun bear; p. 8, tiger, pig-tailed macaque and baby; p. 11, long-tailed macaque, Iban warrior; p. 18, pig-tailed macaque; p. 24, head of sun bear; p. 29, wild pig, bearded pig; p. 82, aquarium tanks; p. 101, common rose butterfly; p. 135, orang utan. **Jansen, Jeffery Mark**, p. 8, long-tailed porcupine; p. 51, Bulwer's pheasant. **Jasmi Abdul**, p. 9, estuarine crocodiles; p. 134, Taman Negara, Sungai Tembeling; p. 135, rhinoceros. **Khang, Peter**, p. 84, mudskipper artwork; p. 102, moon moth; p. 108, ant and termite body structure. **Khoo Soo Ghee**, p. 106, *Trigona* bee, *Trigona* nests, giant honeybee nests; p. 107, *Vespa affinis* hornet nest. **Khor Choon Liang**, p. 116, grasshopper and cricket, grasshopper life cycle; p. 124, malaria transmission; p. 126, roundworm structure; p. 127, tapeworm structure, fluke life cycle. **Kiew Bong Heang**, p. 60, Wallace's flying frog; p. 74, Malayan corrugated frog, red-eyed ground frog, noisy froglet, Malayan jewel frog, tailed tree frog; p. 75, flying frog, horned toad, *Philautus aurifasciatus*, *Rana blythi*. **Lai Jao Sui**, p. 46, yellow-vented bulbul. **Lau, Dennis**, p. 29, hunter, Penan tribesman; p. 30, *parang* hilt; p. 38, *pua kumbu*; p. 63, *parang* hilt, *pua kumbu*; p. 67, monitor lizard of shells; p. 73, carved door; p. 98, *parang* handle and scabbard. **Lee Han Lim**, p. 122, bedbug, louse; p. 123, flea; p. 128, bedbug; p. 129, crab louse, body louse. **Lee Sin Bee**, p. 21, tapir story; p. 30, mouse deer story; p. 72, snake movement; p. 99, leech movement; p. 99, earthworm movement; p. 112, cicada sound mechanism; pp. 122–3, house. **Leh Moi Ung**, p. 48, house

swifts, edible-nest swiftlet; p. 49, birds' nests, scraping and processing nests. **Lim Boo Liat**, p. 24, shattered tree; p. 32, rolled-up pangolin; p. 34, roosting bat; p. 37, flying lemur with legs around trunk; p. 38, Peninsular Malaysia moonrat; p. 39, pencil-tailed tree mouse, wood rat, long-tailed giant rat, grey tree rat; p. 68, flying gecko eggs; p. 69, paradise tree snake; p. 70, common black cobra, banded krait, blue Malayan coral snake, p. 73, short-tailed python. **Mak Joon Wah**, p. 122, roundworm; p. 123, liver fluke; p. 124, oocyst, sporozoite; p. 125, *Brugia malayi*, lymphoedema, filaria worms, elephantiasis, swamp forest, mosquito larvae, filariasis blood survey; p. 126, snail, roundworm egg, whipworm egg, hookworms, cutaneous larva migrans, liver fluke. **Mohd Yunus Mohd Noor**, p. 47, bird-singing competition, practice; p. 50, kite; p. 53, rock pigeons. **Mohd Zakaria-Ismail**, p. 78, *Puntius lateristriga*, *Rasbora elegans*, *Betta pugnax*; p. 80, ikan temegalan, ikan mata merah, ikan tengas, ikan sebarau; p. 81, ikan baung pucuk pisang, ikan tapah bemban, ikan tapah, ikan baung kunyit, *Clarias teijsmanni*, *Mystus bimaculatus*; p. 83, puffer fishes. **Mokhtar Ibrahim,** p. 114, *Cosmopsairria latilinea*, *Dundubia* sp. **Money Museum, Bank Negara**, p. 11, tin ingot, cockerel, wildlife coins; p. 28, seladang coin; p. 30, *kijang* coin; p. 117, grasshopper tin money. **Munan, Heidi**, p. 55, *kenyalang*. **Muzium Negara**, p. 20, elephants with riders. **Natural History Museum Picture Library**, p. 14, sabre-toothed cat. **Natural History Photographic Agency**: Gerard Lacz, p. 21, tapirs. **New Straits Times Press (Malaysia) Berhad**, p. 20, elephant on raft; p. 23, tiger cubs; p. 73, python eggs; p. 96, scorpion man; p. 113, fireflies in tree; p. 120, fogging; p. 134, tiger and cubs. **Ng, Peter K. L.**, p. 86, swamp terrestrial crab, giant torrent prawn, snapping prawn; p. 89, ghost crab, hermit crab; p. 92, mating horseshoe crabs; p. 93, mangrove horseshoe crab. **Nicholas, Colin**, p. 29, Sarawak tribesman; p. 51, Keningau Murut women; p. 96, scorpion with babies. **Ooi, P. A. C.**,

Picture Credits

p. 117, oriental migratory locust. **Phillipps, Karen**, pp. 54–5, hornbills; pp. 66–7, mangrove and freshwater animals. **Photo Bank**: Lucia Tettoni, p. 32, shaman's jacket. **Picture Library**: C. K. Chong, p. 71, Penang Snake Temple, snakes in temple; Christopher Liew, p. 65, Sam Poh Tong temple; Manu Govindasamy, p. 18, silvered leaf monkeys; Ng Phoe Heng, p. 63, crocodiles; Geoffrey Smith, p. 120, flooded plantation; Tan Swee Lian, p. 65, turtles at temple; Arthur Teng, p. 11, pig-tailed macaque and owner. **Pos Malaysia**, p. 9, *hidupan liar* miniature; p. 22, clouded leopard stamps; p. 107, wasp stamp. **Radin Mohd Noh Saleh**, p. 10, bullock cart; p. 47, bird cage; p. 50, quail traps; p. 51, embroidered pillow end; p. 53, quail traps; p. 58, *bangau*; p. 50, walking stick; p. 74, *moyang katak dan kala*; p. 81, storage baskets, fish traps; p. 86, walking stick; p. 125, *moyang kala*. **Ritchie, James**, p. 29, Punan Busang, preparing pig; p. 96, Iban man. **Romo, Steve**, p. 21, woolly rhinoceros. **Sabah Tourism Promotion Corporation**, p. 16, logo. **Science Photo Library**: Tony Brain, p. 129, oriental rat flea; CNRI, p. 123, p. 127, beef tapeworm head, p.129, head louse; Eye of Science, p. 129, *Yersinia pestis* bacteria; Bruce Iverson, p. 126, roundworm eggs (light micrograph), eggs in ovary of roundworm, p. 127, hookworm copulatory bursa; K. H. Kjeldsen, p. 128, cat flea; London School of Hygiene and Tropical Medicine, p. 121, dengue virus; H. Pol/CNRI, p. 131, *Rickettsiae*; Sinclair Stammers, p. 126, clump of whipworms. **Seow, Francis**, p. 60, agamid lizard; p. 61, turtle; p. 64, Burmese brown tortoise, Malayan snail-eating turtle; p. 100, cicada; p. 110, *Phyllium giganteum*, *Tagesoidea nigrofasciata*, *Phyllium bioculatum*; p. 111, *Haaniella echinata*, *Marmesoidea rosea*, *Dajaca monilicornis*, *Phaenopharos struthioneus*. **Shafie bin Haji Hassan**, p. 50, man with quail trap; p. 53, man with bird cage. **Soepadmo, E.,** p. 34, pollinating bat. **Sui Chen Choi**, pp. 12–13, mammals; pp. 52–3, birds of prey; pp. 78–9, freshwater ecosystems; p. 85, mudskippers; p. 101, skipper butterfly and caterpillar; pp. 104–5, dragonflies and damselflies; pp. 132–3, protected animals. **Tan, B. H.**, p. 88, sand-bubbler crab burrow; p. 89, fiddler crab, soldier crab, sand-bubbler crab; p. 91, Chinese horseshoe crab; p. 92, book gills; p. 93, coastal horseshoe crab, Chinese horseshoe crab,

horseshoe crab tail. **Tan Hong Yew**, p. 11, buffalo cannon; p. 23, tiger and other cats; p. 72, nonpoisonous snakes. **Tan, Leo**, p. 92, juvenile horseshoe crab. **Tara Sosrowardoyo**, p. 63, tin money. **Wildlife Art Agency**, pp. 118–19, cockroaches; p. 120, mosquito artwork, mouthparts of mosquito. **World Wide Fund for Nature Malaysia**, p. 36, Prevost's squirrel; p. 39, common treeshrew; p. 40, purple heron; p. 57, rufous piculet; p. 70, Wagler's pit viper; p. 97, pill millipedes; p. 113, *Batocera albofasciata*. Amlir Ayat, p. 103, mounted butterflies; Patrick Andau, p. 28, *banteng*; Azwad M. N., p. 47, common myna, p. 50, great argus pheasant, p. 51, Malaysian peacock pheasant, Javanese green peafowl, p. 52, Nicobar pigeon, p. 53, peaceful doves; Balu Perumal, p. 135, river terrapin, eggs, hatchery; Raleigh Blouch, p. 21, wallowing rhinoceros; David Bowden, p. 59, lesser adjutant; Bruce Bunting, p. 4, clouded leopard; Chew Yen Fook, p. 8, yellow-breasted sunbird, p. 57, common goldenback woodpecker, p. 58, black-crowned night herons, purple heron in flight, p. 59, lesser treeducks, p. 65, male painted terrapin (breeding colour), p. 91, crimson millipede, p. 100, damselfly, p. 104, dragonfly at rest, p. 115, juvenile shield bug; Chin Fah Shin, p. 68, flying lizard, p. 108, brown crazy ants; Gerald Cubitt, p. 8, lime butterfly, p. 27, binturong, p. 60, red-tailed racer, p. 61, red-tailed pit viper, p. 75, pitcher plant, p. 77, fiddler crab, p. 84, mudskipper movement, p. 88, hermit crab, p. 91, painted leech, p. 95, pitcher plant, p. 97, cave centipede, p. 98, painted leech, p. 100, Rajah Brooke's birdwing, p. 102, Malay lacewing, p. 133, muntjak; G. W. H. Davison, p. 35, flying bats; C. M. Francis, p. 35, bats in cave, p. 48, mossy-nest swiftlet, p. 51, Malaysian peacock pheasant egg, p. 57, banded kingfisher; Hans Hazebroeh, p. 27, three-striped palm civet; Fred Hazelhoff, p. 56, common kingfisher; Hymeir Kamaruddin, p. 119, single cave cockroach; Andrew Johns, p. 71, king cobra; Kanda Kumar, p. 47, black-naped oriole; M. Kavanagh, p. 18, slow loris, p. 43, brahminy kite, p. 50, crestless fireback pheasant, p. 109, termite mounds; A. Kemp, p. 195, damselfly; G. Kuehn, p. 39, lesser treeshrew; Rodney Lai, p. 21, tapir tracks, p. 25, sun bear, p. 44, collared scops owl, p. 52, green-winged pigeon, p. 58, milky stork, p. 91, giant millipede, p. 115, lantern fly; D. J. W.

Lane, p. 89, coconut crab; James Loh, p. 10, tapir, p. 44, buffy fish-owl; Y. P. McNeice, p. 52, pink-necked green pigeon; Oon Swee Hock, p. 40, white-bellied sea eagle, purple heron, p. 46, oriental white-eye, p. 47, jungle myna, hill myna, p. 52, pied imperial pigeon, p. 53, spotted dove, p. 56, black-capped kingfisher, p. 58, little heron, purple heron (standing), great egret, little egret, p. 59, cattle egret, common sandpiper; Junaidi Payne, p. 17, orang utan at rehabilitation centre, p. 20, charging elephants, p. 106, *tualang* tree; Cede Prudente, p. 109, *Myrmecodia tuberosus*, p. 114, cicada and cricket; R. Rajanathan, p. 37, East Malaysian moonrat, p. 44, brown wood-owl; Ken Rubeli, p. 21, elephant tracks, p. 37, grey-bellied squirrel, p. 113, *Cyriopalus wallacei*, cockchafer beetle, firefly, p. 118, cave cockroaches; Ken Scriven, p. 25, red dog, pp. 40–1, purple-throated sunbird, p. 44, common scops-owls, reddish scops-owls, p. 45, bay owl, Gould's frogmouth, p. 80, p. 81, fisherman with catch, p. 94, *Liphistius* spider burrows, p. 112, scarab beetle; Dionysius Sharma, p. 27, little civet, common palm civet, p. 45, long-tailed nightjar, p. 65, male painted terrapin (non-breeding colour and intermediate phase), female painted terrapin, painted terrapin eggs, p. 135, releasing river terrapins; S. Sreedharan, p. 10, mouse deer, p. 27, masked palm civet, p. 46, straw-crowned bulbul, magpie robin, white-rumped shama, p. 51, red jungle fowl, p. 52, jambu fruit dove, p. 56, white-collared kingfisher, blue-eared kingfisher, p. 57, buff-necked barred woodpecker, maroon woodpecker, p. 59, purple swamphen, white-breasted waterhen, p. 133, crested wood partridge; Tan Yam Heng, p. 40, white-throated kingfisher; Tengku Nazim, p. 28, *selembu*; T. Whitmore, p. 21, rhinoceros tracks; Edward Wong, p. 82, golden dragon; Frank Yew, p. 91, black scorpion; Dennis Yong, p. 37, flying lemur; Yusof Ghani, p. 28, seladang, p. 98, leech bites; Chris Zuber, p. 44, Malay eagle-owl. **Yap Han Heng**, p. 118, harlequin cockroach; p. 119, housefly, blowfly; p. 120, mosquito larvae; p. 121, *Aedes albopictus*, *Anopheles dirus*, *Culex quinquefasciatus*, *Mansonia uniformis*; p. 122, mosquito; p. 124, *Anopheles maculatus*; p. 125, *Mansonia* mosquito. **Yeap Kok Chien**, p. 16, orang utan nesting; p. 17, orang utan movement; p. 34, bats; p. 35, bat echolocation; p. 63, handbag and shoes. **Yong Hoi Sen**, p. 79, giant Malayan toad, amphibian eggs,

tadpoles, stonefly nymph, *Parapoynx bilinealis*, pond skater; p. 84, mudskipper camouflage; p. 85, mudskipper burrow; p. 86, mountain river crab, *Geosesarma gracillimum*; p. 87, *Pomacea canaliculata*, *Lymnaea rubiginosa*, *Indoplanorbis exustus*, *Robertsiella* snail; p. 90, golden web spider; p. 91, horned spider; p. 94, nursery web spider, jumper spider (male and female), crab spider, horned spider, signature spider, *Liphistius* spider; p. 95, *Nepenthes* crab spider, fighting spider, *Leucauge* spider, wolf spider, golden web spider; p. 97, centipede, millipede; p. 100, *Vestalis amoena*; p. 102, atlas moth; p. 104, *Lathrecista asiatica*, *Orchithemis pulcherrima*; p. 105, *Cratilla metallica*; p. 106, entrance to *Trigona* bee nest, *Apis cerana* bees; p. 107, wasp colony, *Vespa affinis* hornet; p. 113, net-wing beetles, leaf-cutting weevil, two-horn rhinoceros beetle, rhinoceros beetle, tortoise beetle; p. 116, *pelesit* grasshopper, *Tauchira polychroa*, *Nisitrus* sp.; p. 117, grasshopper nymphs, *Phyllomimus* sp., *Chlorotypus gallinaceus*, *Oxya hyla*, *Valanga nigricornis*; p. 118, cockroach egg case; p. 119, forest cockroaches. **Yu-Chee Chong Fine Art, London**, p. 39, bamboo rat.

Slow loris (*Nycticebus coucang*)